The Genesis of Nature and the Nature of *Genesis*

A Schoolmaster looks at Evolution and Creation

Norman Parker

Published by New Generation Publishing in 2015

Copyright © Norman Parker 2015

First Edition

The author asserts the moral right under the Copyright, Designs and Patents Act 1988 to be identified as the author of this work.

All Rights reserved. No part of this publication may be reproduced, stored in a retrieval system or transmitted, in any form or by any means without the prior consent of the author, nor be otherwise circulated in any form of binding or cover other than that which it is published and without a similar condition being imposed on the subsequent purchaser.

www.newgeneration-publishing.com

"Religion and natural science are fighting a joint battle in an incessant, never relaxing crusade against scepticism and against dogmatism, against disbelief and against superstition..."

Max Planck, *A Scientific Autobiography*

Contents

Foreword i
Introduction 1

Part 1 - The Genesis of Nature
Chapter 1 The Great Debate 13
Chapter 2 The Evidence for Evolution 25
Chapter 3 Mechanisms of Evolution 38
Chapter 4 Macroevolution and the neo-Darwinian
           Synthesis... 48
Chapter 5 From Biology to Everything 62
Chapter 6 The Characteristics & Origin of Life 71
Chapter 7 Evolutionary Developmental Biology (Evo
           devo).. 81
Chapter 8 Innate Transformation.................................... 88
Chapter 9 Stories and Just So stories 107

Part 2 - The Nature of Genesis
Introduction to Part 2.. 117
Chapter 10 The Two Books.. 120
Chapter 11 Christian parts of the spectrum 131
Chapter 12 Fossils 149
Chapter 13 Gossery; Criticisms of *Omphalos* 160
Chapter 14 The Book of God's Word 168
Chapter 15 Comments (1): Consilience; What is it
            about Darwin?; Tit for Tat arguments;
            The words "evolution" and "natural
            selection"... 183
Chapter 16 Comments (2): Science and religion;
            Mind and matter; Discontinuities in
            Creation; Variations on a theme................ 198
Chapter 17 Easing the Situation 212
Chapter 18 Advice to Students 220
Chapter 19 The genesis of Nature and the nature of
            Genesis.. 225
Glossary.. 231
Bibliography.. 235

Foreword

A Christian acquaintance named Dave once asked me to recommend a book on the evolution / creation debate. I replied that I could not, because all the books that I had read were very one-sided. Creationist authors, arguing against evolution, gave only the facts which supported their own case. Evolutionist authors did the same for *their* case. So I had to tell him that I knew of no book that dealt with the subject in a balanced way *. "Then why don't you write one?" he asked me.

My experience of this topic is of teaching and explaining evolution, mainly to school Sixth-formers. You may wish to read something written by an expert. But, then, which expert will you choose? There are working scientists, many with doctorates, on both sides of the dispute, not to mention theologians and philosophers. Evolution is too wide a subject for anyone to be an expert on more than part of it.

I mention Millfield School a lot in Part 1 because that is where I taught biology for the last thirty or so years of my teaching career. Ours was a buoyant department run, with enthusiasm and refreshing good humour, by my colleague Mike Cole. The Science Staff at any one time included eight or nine physicists, the same number of chemists and eleven or twelve biologists. We shared scientific problems and bounced ideas off each other. We had the use of large departmental libraries and kept reasonably up-to-date with new books as they came out and such magazines as *New Scientist* and *Scientific American*.

I owe my colleagues a large debt for their kindness and wisdom. I should like to make it clear that any mistakes or misguided opinions in this book are mine alone and not shared by any of my fellow-teachers.

There is just one claim which I might make: I am one of those who have actually read both the Bible and *The Origin of Species* from cover to cover. There cannot be

i

very many of us. Perhaps we ought to form a society and wear a special tie.

Part 1 (The Genesis of Nature) is "scientific" in the narrow sense that most of the facts and ideas are topics that might be discussed in a science lab. (Mention of the Bible or religious belief is in passing.) Part 2 (The Nature of *Genesis*) ranges more widely and anything goes.

....

* Since my conversation with Dave at least two books have been published which I would have recommended if they had been available then, The two are *Responding to the Challenge of Evolution* by Kevin Logan, and *God's Undertaker* by John Lennox.

(for details see the Bibliography)

Part 1 The Genesis of Nature

Introduction

What is it about Darwin ?

Every few years a book is written with a title such as "*Where Darwin Went Wrong*" or "*Evolution: The Discredited Theory*" and there is a flurry of interest for a few days. Creationists are delighted and buy copies for their friends. Popular evolutionists write angry reviews for the Science page of newspapers. There may be some letters in the *New Scientist*, perhaps one saying, "It's about time someone had the courage to write such a book!" but most saying, "Oh no ! The religious obscurantists are knocking Darwin again!"

When such a book appeared a few years ago a reviewer in a national newspaper asked "What is it about Darwin ?" He was puzzled because, after more than a century of scientific research supporting the theory of evolution, seemingly intelligent people still couldn't accept it.

That phrase stuck in my mind: "*What is it about Darwin?*" There is obviously something special about Darwin and his theory. You don't get book after book attacking the quantum theory, or the arrangement of the elements in the Periodic Table. Surely the evidence for a scientific theory devised a century and a half ago would be reasonably well known, and accepted, by now?

So what is it about Darwin ?

Historical Science

The most important thing to say about Darwin's theory of evolution is that it refers to what happened in the past, mostly the remote past. The sciences which deal with the past are different in several important ways from present-day science. The techniques are different and most of the results cannot be established with certainty. The late Stephen Jay Gould - a leading palaeontologist and author of many books on evolution - divided science into the "hard sciences" (physics, chemistry, animal physiology and so on) and the "historical sciences" (cosmology and evolution). Statements from the hard sciences mostly use the present tense: "Copper sulphate *forms* a blue solution in water"; "pollen grains *develop* inside the anther of the flower." Historical science statements are in the past tense: "In the Carboniferous era amphibians *gave rise* to reptiles"; "cynodonts *evolved* a mammal-type lower jaw."

There is disagreement about the status of historical science. Is it as reliable as hard science? You can perform an experiment to demonstrate that plants give off oxygen in sunlight, but you cannot do an experiment to show how birds evolved. It will be pointed out that science has several different techniques that it can use to establish facts and experimentation is only one of them. Nevertheless there is a difference in the *certainty* of the results obtained by the two sorts of science: you cannot go back to the past to check your detective work, however ingenious. This is why people continue to quarrel about evolution and cosmology, but not about optics or chemistry. It explains why some people get so angry about evolution: it is terribly frustrating when folk don't accept your story of what happened in the past and there is no way you can *prove* them wrong.

Some scientists strongly resent the implication that historical science is not as well established as hard science. But the two are different. Here are some facts with a bearing on this question:

1. In our biology department we had a variety of textbooks, English and American. In the Introduction to one textbook is the statement: "All competent biologists now accept evolution." Another book has this sentence in its Foreword: "No serious biologist today doubts the fact of evolution." But you never find the statement, "No serious biologist today doubts the fact of photosynthesis." The textbooks of other sciences do not have equivalent statements. Physics books do not start with, "All

competent physicists now accept Ohm's Law." It could even be suggested (mischievously but reasonably) that the fact that such a statement is made at all probably shows that it is not true.

2. I heard someone described recently as a "staunch Darwinist." You don't hear of a staunch geneticist, or a staunch biochemist. Staunchness goes either with historical science or with very far-fetched ideas in hard science. And can a scientist be open-minded and staunch at the same time?

3. Every now and then some idea (such as Intelligent Design theory) is labelled "an attack on science." Even if it is an attack on science (it rarely is) as far as I know this is never an attack on hard science.

4. In many an article describing the discovery of a new fossil, you will read that "all the textbooks will have to be re-written." This means that the textbooks must have been wrong up to that point. It is no one's *fault* that they are wrong, but they were wrong nevertheless. Now if some part of a book is shown to be wrong, might it not affect our confidence in other parts? How do we know that the new ideas won't be superseded or debunked in their turn? This sort of thing might happen in any science, of course, but it happens far more often in a historical science. Do not misunderstand me here. It is one of the strengths of science that it is self-correcting; new discoveries correct earlier mistakes. But this need for correction leaves no room for arrogant assertion. Theories about the distant past must be held provisionally.

Some people would say that this is trivial and others that it's important. One thing is certain: historical science is more likely to need revision than hard science.

These facts have a bearing on arguments between creationists and their opponents. In book after book evolutionist authors have ridiculed creationists by likening them to flat-earthists, or have stated that "evolution is as well established as the fact that the earth orbits the sun." But these arguments are based on false analogies: the shape and orbit of the earth are part of hard science, but evolution is not. You can go out next Tuesday and take measurements and photographs to establish that the earth is round or that it moves round the sun, but you cannot perform equivalent experiments on events in the past. This explains a curious fact: every astronomer in the world would agree about the shape and orbit of the earth, but many hundreds, probably thousands, of biologists and other scientists do not accept Darwinian evolution.

Another argument is this: "Science has given us aeroplanes and medicine. Therefore science works. Evolution is also part of science; therefore, because planes and medicine are reliable, evolution is true." Did you spot the sideways slip from hard to historical science? Please note: this does *not* mean that evolution is therefore a wrong idea, but false analogies with hard science do not help to support it. An even worse argument, often implied, is this: "Science has given us aeroplanes and medicine. Therefore science works. I am a scientist. Therefore what I tell you about evolution is true."

This is why - as far as the *science* is concerned - it is unwise to be dogmatic. When something cannot be proved, it is pointless to call a particular position either right or wrong. New facts may turn up. New interpretations of 'old' facts may be devised. No one knows what happened in the past; or, if they think they know, they simply cannot demonstrate it. This particularly applies to a *process*, like natural selection, because it cannot be fossilised. The most you can do is provide a "best interpretation" in the certain knowledge that some other people, equally knowledgeable and as clever as you, will not accept it. (Note the proviso: as far as *science* is concerned. Beliefs and philosophy are another matter entirely.)

Other problems with Darwinian evolution

A second point about Darwin's theory is that it is counter-intuitive. It is not only creationists who find something preposterous in the idea that an empty planet can, "on its own", become full of countless thousands of living things, or the belief that consciousness and intelligence are the products of chemical reactions. I saw an anti-evolution cartoon the other day in which a professor had written on a blackboard, "Hydrogen is a colourless, odourless gas, which, given enough time, turns into people." Quarrel with that as much as you want, but it's hard not to feel some sympathy to the view behind it. Another example of a counter-intuitive idea is the implied claim that *information* can somehow build up "from scratch." Information is embedded in, or carried by, the genetic code, and many discussions about evolution end up with arguments about the source of this information: how can information arise and build up meaningfully in the absence of some form of intelligent input?

Thirdly, Darwin's theory, in its modern form, is not just a matter of simple biology. It appears to give an answer to deep questions, such as "What is Man?" and "Have our lives any meaning or purpose?" Are these just *biological* questions? It is not only creationists who believe that there are better answers to such questions than those derived from evolutionary theory.

Fourthly, following on from the last point, evolution has in the last half-century become linked to a strident and aggressive atheism which is very different in tone from Darwin's original conception. A reasonable biological theory is used to underpin a contentious worldview. Darwinists often complain about the harm that creationism does to "science." The hijacking of evolution to support an ideology like atheism does at least as much harm to science. For a materialist, Darwinism simply *must* be right, whatever the evidence. That is why ultra-Darwinists are so fierce when anyone questions evolution. But even without the explicit atheism, this aggressive insistence that Darwinism *must* be right is a deeply unscientific attitude. James Le Fanu put his finger on it in a recent book: "The greatest obstacle to scientific progress, after all, is not ignorance, but the illusion of knowledge." [1]

It is not surprising that many people see Darwinism nowadays more as a worldview like Marxism than as part of science. For it is not just a matter of evidence, in spite of what is often claimed. The biological facts are available to everyone, but what you make of them - how you use this evidence - depends on your philosophy or beliefs. It is possible for someone to accept evolution without becoming an *evolutionist* and to accept Creation without being a *creationist*.

Some Principles

Origins is a huge topic. Even a "popular" book like *What Evolution is* by Ernst Mayr has two hundred references in its Bibliography, and that includes no creationist books at all. There are over fifty entries in my own Bibliography at the end of this book (and, yes, I have read them all) and I have read or dipped into many other books which I did not list. There are countless articles and television and radio programmes (especially recently because of Darwin's bicentenary in 2009) not to mention the sprawling mass of information on the Internet, much of it unchecked and unreliable. Evolution covers so much ground, and is so detailed, that it is hard to think straight about it. So here are some principles which I have found useful.

Check the authority

Evolution is studied under many different headings. Darwin himself was a naturalist, and evolution is primarily a theory in biology, but the modern concept has input from geology, biochemistry, mathematics, cosmology, theology and philosophy. It is impossible for anyone to be an expert in all of these, and an expert in one subject may be an amateur in others. A philosopher may be hopeless at maths and a zoologist may be completely ignorant about theology.

It is a useful principle to trust an author (at least provisionally) when he is writing about the facts of his own subject, but to be very wary when he is writing about anything else.

People who are writing about difficult topics that they have not studied in depth are likely to make mistakes. Here are two examples. The first comes from a small book which attacks evolution, written by a theologian. The Biblical section is fine, but the science is all over the place. In a section on biochemistry is this extraordinary statement:

"To survive, a DNA chain must be made up of hundreds of *"pure" right-handed amino acids* (capable of bonding to a different chain of *"pure" left-handed nucleotides* - protein enzymes)."

An A level biology student should be able to pick out at least seven mistakes in that single sentence. (I make it ten.) After reading that, how can the reader have any confidence in the rest of what the author tells us about biology? He does not understand what he is writing about. The second example is from a book written by an evolutionist:

"faith . . . means blind trust, in the absence of evidence, even in the teeth of evidence."

Here there is one big error rather than many small ones. The statement is simply false. I have known hundreds of Christians, and for not one of them does faith mean "blind trust." They believe what they do because they have, or claim they have, very good reasons. Some of them are not very articulate about it, but those who are will give you plenty of evidence for their beliefs.

As far as possible, stick to facts

"Origins" is a controversial subject and most authors want you to share their beliefs. Beware of opinions, especially aggressive opinions. Always go back to the actual facts. This is harder than it sounds because the word *fact* can mean different things. "Evolution is a fact!" is a common phrase, but consider this chain of statements:

The objects I am looking at are fossil skeletons
They have a common pattern, so can be compared and classified
This shows that there is a relationship between them
This can only be an evolutionary relationship
This evolution must be the result of Darwinian natural selection

Are the statements equally factual?

In the same way, if the conversation turns to creationism and someone says "the Bible tells you so-and-so" go and read the passage for yourself. You may be surprised. Beware of sweeping statements. "All scientists now accept . . . " may simply mean "This is what I think, and surely everyone else agrees." "The church in the twenty-first century no longer believes in . . ." may just mean "I have never believed in . . ." Don't be misled by appeals to prestige. Evolutionist writers often refer to other evolutionists as "great" or "brilliant." Creationists refer to each other as "acclaimed author" or "noted scholar." Even if all these brilliant people were really as clever as implied, don't forget that great men can still write great nonsense.

I have a friend, Mark, who is a "theistic evolutionist" (i.e. he accepts Darwinian evolution but is also a Christian.) I remember telling him something factual - about dinosaur bones - that I had read in a book by a creationist author, and he replied, "Ah, but where's he coming from?" This had the effect of diverting our attention from the dinosaurs to the problem of how far to trust authors who are *coming from* a particular position, and we had quite an argument about it. We laugh about our argument now, but it does raise some important points. Most authors in the origins debate hold very strong views and want to persuade their readers to share them. So, being human, they will emphasise the facts that support their own case and leave out any facts that might undermine it. Young-earth creationists list all the facts which support the view that the world is only a few thousand years old, such as the amount of salt in the sea. Evolutionists list all the facts that show that the world is billions of years old, such as the ratios of radioactive elements in certain rocks.

Is it wrong that these writers are "coming from" their particular positions? No, for two reasons. Firstly, everyone is coming from somewhere, and you would be scarcely motivated to write a book at all unless you felt fairly strongly about it. Secondly, that is how science works: facts, in order to make sense, must be tested against a hypothesis.[2] Charles Darwin wrote, "How odd it is that anyone should not see that all observation must be for or against some view if it is to be of any service!" [3]

But you have to be careful. Here is a statement by Pierre Grassé, an eminent French biologist:

"Biochemists and biologists who adhere blindly to the Darwinism theory search for results that will be in agreement with their theories and consequently orient their research in a given direction . . . This intrusion of theories has unfortunate results: it deprives observations and experiments of their objectivity . . ." [4]

The key word is *blindly*. Everything will look pink to someone wearing rose-tinted glasses. J B S Haldane, was a famous scientist and mathematician who wrote about evolution in the 1940s: his writing reads very strangely nowadays because he saw all his science through Marxist spectacles.

The "Nuffield Filter"

Many years ago, at Millfield, we adopted the Nuffield syllabus for some of our biology classes. This was a practical-based syllabus which emphasised a *scientific* approach to the subject. Pupils were encouraged to think for themselves, to make hypotheses, to devise experiments to test the hypotheses, to evaluate the results, and so on. Often a statement would be made in the Nuffield text or on a test-paper, and the pupils would be asked to think about it critically according to these questions:

Is this certainly true?
Is this probably true?
Is this possibly true?
Is this false?

I call this the Nuffield Filter. (You can think of it as a sort of intellectual sieve .) It is useful to apply it to *assertions* made by authors or speakers. Because evolution is such a contentious topic, there is a tendency for people to make sweeping statements to back up their own entrenched position. Have a look at these statements, all taken from various writings on evolution, and try applying the filter to them:

a) "Evolution is as firmly based a science as, say, astronomy or parasitology"
b) "If you meet somebody who claims not to believe in evolution, that person is ignorant, stupid or insane..."
c) "Evolution is neither testable nor observable, and therefore is not science."
d) "An attack on evolution is an attack on all of science."
e) "The scientific method is the only source of knowledge."

f) "Natural selection is the only effective agency for producing change in biological evolution."

One's first reaction is to think of these statements as true or false (even "sensible" or "silly'") but applying the Nuffield filter gives a more measured response. You will probably guess that I have chosen these examples because I think that none of them is true. Am I right?

Newton's Dog.

In 1860 Charles Darwin wrote to his friend Asa Gray, about the relationship between evolution and religion:

"I am inclined to look at everything as resulting from designed laws, with the details, whether good or bad, left to the working out of what we may call chance. Not that this notion at all satisfies me. I feel most deeply that the whole subject is too profound for the human intellect. A dog might as well speculate on the mind of Newton. Let each man hope and believe what he can." [5]

When I was teaching biology I learned to be able to answer "I don't know" to many of my pupils' questions. In the origins debate there are many things that are simply unknown, in spite of the brash assertions of some of the debaters. But there are also "Newton's dog" questions with unknowable answers; about matters beyond human understanding, and about which it would be foolish to hold opinions.

"What is it about Darwin?" Unlike most other biological theories, Darwinism has changed from being a science to being a philosophy. From everything that I have read by him, or about him, I believe that Darwin himself would have been most unhappy about this state of affairs. Another Nineteenth Century thinker, often compared with Darwin, is Karl Marx. Marx is said to have remarked, "All I know is that I am not a Marxist." Perhaps Darwin, given the chance, would likewise say, "All I know is that I am not a Darwinist."

Notes to Introduction

1. *Why Us ?* (Chapter 4) by James Le Fanu

2. See "Colorful Pebbles and Darwin's Dictum" by Michael Shermer from *Scientific American* April 2001.

3. from a letter that Charles Darwin wrote to a friend on September 18, 1861

4. *Evolution of Living Organisms* 1977

5. Asa Gray was a Harvard botanist and a Christian evolutionist. The quotation is from a letter to Gray dated May 22 1860. Darwin wrote "I am bewildered. I had no intention to write atheistically." But he was much troubled by the problem of reconciling the idea of "a beneficent and omnipotent God" with the evil and misery in the world.

Chapter 1

The Great Debate

Some years ago Vernon Blackmore and Andrew Page produced *Evolution: The Great Debate,* written from a historical point of view and packed with information and portraits of the various scientists involved [1]. The "Great Debate" of the title referred to the controversy between evolutionists and their opponents, mostly those known as creationists. There are lesser debates within evolutionism itself, as we'll see later on, but the Great Debate is between those who believe that evolution is the explanation of the variety of living things (including our own very existence) and those who deny it. This debate has taken very different forms on the two sides of the Atlantic. It has been strongly polarised in the United States, particularly since the famous Scopes trial in 1925. The debate there isn't just about biology and religious belief but is also a matter of politics, law and education policy. Nor is it just an intellectual discussion; it is an angry quarrel, almost a war, with the two sides being extremely rude about each other. Atheists attack creationists for their "falsehoods"; one has even described the teaching of creationism as "child abuse." And creationists can be equally fierce. In the first few minutes of a recent American creationist DVD series, the speaker accused evolutionists of "hoodwinking" people about science, and of "brainwashing" school-children.

On this side of the Atlantic the Great Debate has been, until recently, in a much lower key. I have taught evolution, as part of biology, in a number of schools for over forty years, and it was never a controversial topic in the classroom. As a member of various churches and Christian house groups I can record that evolution was rarely mentioned. But during the last half century the situation here has begun to resemble that in America. It is a three-fold change. Firstly, the general public has become much more aware of the debate: creationism is referred to in newspaper articles and on television, mostly by opponents who do not realise that they are providing welcome publicity to the very thing they are attacking. Secondly, there are far more strands to the debate than before. There are now various groupings within evolutionism, creationist societies have multiplied, and there is a wide spectrum of views (see Chapter 10). Thirdly, this debate has begun to appear in school textbooks.

As I have been involved in natural history and biology for most of my life, I can best describe these changes from my own experience. For nearly sixty years I have had to do with the theory of evolution, as a schoolboy, as a student reading natural sciences and as a biology teacher.

Learning and teaching evolution: evolution at school

We were taught biology at school (Monkton Combe, near Bath) by an excellent naturalist named Douglas Brown. Back in the early 1950s there was little else in our timetable besides our three A level science subjects. In the biology syllabus there was very little microbiology or biochemistry and nothing about DNA, which we had not even heard of (The famous paper on the structure of DNA by Watson and Crick was published in 1953, my last year at school). We did plenty of fieldwork in the rich Somerset countryside but, mercifully, "coursework" had not yet been devised. Mr. Brown always "taught from the specimen" and by the end of their A level courses his pupils had acquired a wide factual knowledge. For me the glue that held all this knowledge together was not the idea of evolution, i.e. something that had happened in the past, but the awareness of the *present* similarities, differences and relationships between living things. We compared the cells of algae, ferns and flowering plants and the anatomies of beetles, wasps and ants. We ourselves prepared, and then compared, the skeletons of a frog and a rabbit. We dissected the blood systems of earthworms, frogs and rabbits, and had added to these the arterial systems of the guinea pig, the dogfish, the skate and - on one memorable occasion - a fairly large shark. These comparisons and relationships were fascinating, but I cannot remember being much interested in how they came to be.

Mr. Brown taught evolution in a matter-of-fact, scientific way. Some of his pupils were "creationists" (though the word was not used then) and argued against evolution, but I cannot remember his being much annoyed by this. Monkton Combe is a Christian foundation (when I was there its ethos was a cheerful and non-pushy low-church Anglicanism) but there was very little conflict about evolution, which was simply taught as part of the syllabus.

In the early 1950s I was largely unaware of the wider controversy about evolution in the world outside school. The only anti-evolution organisation I'd heard of was the Evolution Protest Movement. Like the creationist groups today, the EPM included many scientists. Most of the EPM literature dealt with biological topics such as vestigial organs or "missing links" rather than the age of the earth. [2] EPM authors were not particular about how to interpret Genesis, and I don't remember any reference to the six days of Creation. There was a strong emphasis on the links between Darwinism and the evils of Nazism and Marxism. Some leaflets simply attacked atheists or "Freethinkers." They were particularly angry with the BBC which was seen as heavily biased towards evolution and intolerant of any other views.

I cannot remember any equivalent Evolutionist writings. I realise now that there were such books around, written by J B S Haldane, Julian Huxley, and others, but in my schooldays I never came across them. I have read a number of these books since and it is clear that the argument then was more rationalist-versus-religious than evolutionist-versus-creationist.

Evolution at University

I studied Natural Sciences at Cambridge, including two years of botany, zoology and geology and a year of vertebrate palaeontology. Organic evolution was there in the background as the unifying concept throughout. We studied invertebrate fossils in the geology department and vertebrate fossils in the zoology department. We were thus studying evolution, in great detail, for month after month, year after year. But little was taught about neo-Darwinism or natural selection and the Origins debate was never mentioned.

We were taught by many eminent scientists. I had no idea what philosophical ideas or religious beliefs these professors and supervisors held, for the very good reason that they didn't tell us - it having nothing whatever to do with science. I think that this reticence, as well as being scientific, was also wise. The great virtue of the teaching we were given was that, as well as giving us knowledge and skills, it enabled us to begin to distinguish between wisdom and mere cleverness. I learned what a leading scientist should be like: thoughtful, not arrogant; measured, open-minded and ready to listen carefully to contrary views.

Most of our teachers were very cautious in their conclusions, and they were careful to pass this caution on to their students. This reluctance to *assert* anything was almost comic in the case of some of our geology lecturers. You could never get them to tell you what a piece of rock *was*, even though they knew perfectly well. They would only discuss ways of finding out. If you happened to be, at the time, in a quarry near Nuneaton or on a mountainside in Arran this could prove frustrating. The equipment you needed was usually miles away.

All this scientific teaching was within an evolutionary framework. Any problems that I had had about reconciling evolution with creation were long since forgotten, and I was very impressed with the matter-of-fact way in which evolutionary ideas were dealt with in all the various disciplines. This wise way of dealing with organic evolution contrasts strongly with the arrogance of some modern evolutionary authors.

At one point I found a booklet from the Evolution Protest Movement in our college common-room. As well as attacking evolution in a rather simplistic way, it contained many factual mistakes. I wrote to the EPM, explained about the mistakes, and suggested that the booklet was counter-productive: more likely to put students off Christianity than to convince them that evolution was wrong, especially as they were confusing evolution with evolutionism. I had a courteous reply from the secretary of the group, and was put in touch with the EPM President, Captain Bernard Acworth, who thanked me for what I had written and said he would correct the errors in the next edition of the booklet. This was the start of a long and friendly correspondence between us. We never came to any real agreement about the science, and I must have represented many of the things he hated, but this did not affect our friendship or mutual regard in any way.

I graduated in the summer of 1959. I realised the significance of the date when I was half way through writing this chapter. For this was the year when the centenary of the publication of Darwin's *Origin of Species* was celebrated: a turning point in the Great Debate, when the attention of the world was directed to Darwin, and his theory was explicitly linked to atheism in lecture after lecture. Perhaps the fact that our particular degree courses had finished just before the Darwin centenary explains how we managed to go through three years of Natural Sciences without hearing a word from any of our lecturers about evolutionism.

Teaching evolution

In none of the four schools in which I taught biology was the teaching of evolution a matter of controversy. None of my fellow biology teachers was a creationist, in the current meaning of that word. However in one school one of the RE teachers, a retired clergyman, was well-known for his strong anti-evolution views. He was a friendly and knowledgable man, but the pupils knew that if they wanted a work-free lesson, they had only to mention evolution. This was as a red rag to a bull: my colleague would drop everything and respond vigorously until the bell went. The head of biology couldn't understand why his worst set, a group of almost unteachable students, suddenly showed an enthusiasm for evolutionary theory, asking questions and taking notes. He realised later that they were collecting ammunition for their next RE lesson. (I make no comment on the RE teacher's views, but must point out that his classroom technique wasn't very good!)

The biology department at Millfield was large, with anything between ten and twelve biology tutors at any one time, most of whom taught A level. This was an excellent teaching environment, because we could learn from each other, discuss new books, and so on. We turned an unexpected roof space into a large display room and once a year we put up an extensive Evolution Display with many wall-charts, models and specimens. This was fun to do and I think that the pupils found it rewarding. We had plenty of fossils (many collected ourselves on field trips), a dubious stuffed kiwi with plastic legs, and an aquarium with a placid axolotl in it for when we taught neoteny. We had living carnivorous plants as examples of specialisation, cases of butterfly *models* and *mimics*, and a rather splendid museum piece over two metres high with casts of the series of horse fossil leg bones. We even had speckled and black (melanic) peppered moths fastened to a branch, with a pair of stuffed nuthatches staring at them. Whether our teaching was successful or not, at least our pupils have looked at a real peppered moth. This famous moth features in all the textbooks but I wonder how many students of evolution have actually seen one. We encouraged our pupils to make their own notes on evolution from the display and textbooks and then went through the topic, helping them to sort out the multitudes of facts.

Comments on the teaching of evolution

Evolution as a difficult topic

Even though the 6th form pupils had done three years of GCSE biology and about a year of A level, they found evolution hard to get their heads round. They seemed to understand each idea or piece of evidence as it came along, but were unsure quite how it all fitted together. They muddled all those "*-ation*" words (variation, speciation, adaptation, radiation, isolation, recapitulation....). They found it difficult to distinguish between *evidence* and *conclusions*. Their written work would betray unsuspected muddles: writing answers which suggested that an animal could somehow "evolve" an organ just because it needed it, or that the dark (melanic) form of the peppered moth somehow *knew* that its wings were black, and therefore chose darker bark to settle on. [3]

Our pupils mostly did not do well in "the evolution question" (there was usually one in each exam). A candidate capable of reaching 75% in a question on pollination mechanisms, or DNA chemistry, would get only 20% on evolution. This meant that a good scholar who chose the evolution question did worse than his or her fellows who chose more descriptive questions, and this badly skewed the overall results. In the end, with regret, we had to warn our exam candidates not to attempt the evolution question, purely as a matter of exam technique.

This experience suggests two things about the teaching of evolution:

Firstly, clever A level pupils find evolution difficult. If it is hard for them, what is the likelihood that ordinary people, non-specialists without a scientific background, can understand it? My experience is that most people are muddled about evolution, in spite of the natural history programmes on television, which mostly assume evolution rather than explain it. These sound like sweeping statements. But I have been having conversations about this topic for most of my life, and particularly in the last eight years, and it is rare to find someone with a good understanding of evolution.

Secondly, it is a mistake to introduce evolution too soon into biology courses. GCSE level is much too soon. If A level pupils find the subject difficult, GCSE youngsters will not understand it. This is particularly true nowadays because biology courses have been emptied of much of their factual content. Evolution is a good example of a subject in which a little learning is a dangerous thing. It is better to be ignorant than to have muddled or wrong ideas.

Though now retired, I often visit my old biology department and recently saw a copy of one of the current GCSE biology text-books. There are 50 pages of human biology to start with, followed by 50 pages on evolution, including DNA and cloning. Cell biology comes later on and genetics later still! How can the poor pupils possibly understand evolution if it is taught before they know how a cell works or before they know about heredity? It gets worse: a biology teacher told me that evolution is taught in Year 10 but cell biology not until Year 11 (when the weaker pupils have dropped biology). What sort of understanding of evolution can children possibly have, if they do not know what living things are made of or what cells are? The child's head will be full of concepts, unbacked by factual knowledge, that he or she scarcely understands. I recently read that some experts are urging the government to make the teaching of evolution compulsory in all *primary* schools, on the grounds that

"Evolution is the most important idea underlying biological science. It is a key concept that children should be introduced to at an early stage." You will guess from what I have written above that I think that this is an absurd idea. These experts may know a lot about biology but they cannot know much about teaching children. There are plenty of "key concepts" that are important ideas (general relativity, say, or Marxist economics) but you would not dream of introducing them to small children. I suspect that "introducing children to evolution" is less a matter of science teaching, more one of putting across a world-view (see Chapter 5).

The universality of evolution

It has been claimed that before Darwin the study of biology was tantamount to stamp-collecting, and that it is only since the appearance of *The Origin of Species* that the study of living things made universal sense. Dobzhansky said that nothing in biology makes sense, except in the light of evolution. The Medawars told us that for a biologist the alternative to thinking in evolutionary terms is not to think at all. How true are these statements? Biology did not suddenly start in 1859 and it is unlikely that medical doctors or agricultural scientists think about evolution very much. This is not to disparage evolution, but to try to keep it in proportion as a biological theory. When we did biology, at university or in the classroom, the idea of evolution was not continually in our minds. When you are actually studying living organisms, immediate events and processes loom much larger than historical ones. Although I accepted evolution as a unifying background idea, I was rarely conscious of it, even when we were studying the classification. Nature presented us with a thousand variations-on-a-theme, which were endlessly fascinating in their own right. The 'horizontal' aspects of the classification (distribution, ecology and physiology of the organisms) were normally in our minds, rather than the 'vertical' ones (how did these things evolve?) Our best textbook was subtitled "*A functional approach*" and the main questions before us were "How is this organ constructed?", "How does it work?" and not "Where did it come from?"' Some books on the origins debate state that pupils who study biology at a school which does not teach evolution (usually in America) miss out terribly. This is an exaggeration. I do not think that most of our pupils would have lost very much if we had left evolution out completely.

The wider importance of the topic

At the end of the evolution part of the biology course there was usually some spare time in which to discuss how it related to philosophy or religion. Over the years I discovered several things:

1. The Origins debate did not arouse much interest amongst pupils. There was some dutiful discussion, but no great enthusiasm or strong argumentation. This seemed not to depend on the religious background of the pupils. Millfield is a secular, multicultural, school, and perhaps you might not expect any particularly strong views. But I also had the privilege of teaching biology for a short time at Hebron School in Ootacamund, high up in the Nilgiri Hills of South India. Hebron is a Christian school, originally set up to educate the children of missionaries. All the staff, and most of the pupils, were Christians, and the ethos of the school is based on the Bible, so I wondered whether there would be any opposition to my teaching evolution to the sixth form. However none of the pupils had any problems with the topic and no one seemed particularly interested in discussing it.

At Millfield the biology staff recommended that pupils should take books out of either the department library or the main school library in order to "read around" the topic, but very few ever did. Even the "Scholarship pupils" rarely took a book out of the library. (Recently I went into the school library to borrow a book on evolution. Curious to know how often it had been borrowed from the library, I looked inside the front cover. It had only been taken out twice *ever* - both times by me!)

2. My pupils, being teenagers, tended to take the opposite position to mine. If I happened to argue strongly for evolution, they would find all sorts of counter-arguments. If I mentioned any creationist arguments they instantly became enthusiastic Darwinians. For perhaps most of them it was an intellectual game. Looking back, I'm glad I wasn't dogmatic; they would simply have ignored me.

3. I discovered, not surprisingly, that most pupils thought that the sensible thing was to accept evolution as the scientific view and the Genesis account as the popular view; in the same way that people speak of "the earth orbiting the sun", or of "the sun rising in the morning" depending on the context. This was without any prompting from me. Discrepancies in the two accounts did not bother them in the least. As G K Chesterton wrote:

"(The ordinary man) . . . has always cared more for truth than for consistency. If he saw two truths that seemed to contradict each other, he would take the two truths and the contradiction along with them." [4]

This is a sensible point of view. There is nothing wrong with holding two contradictory ideas in tension until you have enough facts to clear the matter up.

How the Debate has changed

The nature of the debate has changed during my lifetime. As the years went by I became aware that in the wider world evolution was becoming more of a contentious topic. Creationist groups were holding meetings and distributing literature. Articles and letters in Scientific magazines were becoming explicitly anti-creationist. Popular books on evolution, instead of being philosophically neutral, as good science should be, were written from an atheist standpoint. But this discord did not get into our classrooms or laboratories. The pupils were just not interested, and the biology tutors either did not know about these quarrels or had too much sense to raise them in class.

Notes to Chapter 1

1. *Evolution - The Great Debate* by Vernon Blackmore and Andrew Page, Lion

2. Many Christians at this time accepted a view (the "Gap Theory") which attempted to reconcile Genesis with scientific discovery. In the Scofield Bible, popular for several decades at the beginning of the 20th Century, Chapter 1 of Genesis is headed BC 4004, but Dr. Scofield makes room for ""geological time." Here is the footnote for Verse 2:
"But three creative acts of God are recorded in this chapter: (1) the heavens and the earth, v.1; (2) animal life, v.21; and (3) human life, vs. 26, 27. The first creative act refers to the dateless past, and gives scope for all the geologic ages."
Creationists scorn this "Gap Theory" nowadays, but there was a lot of wisdom in it. It certainly caused much less controversy than later interpretations.

3. Peppered moths do not rest on bark anyway; they hide amongst the vegetation. The whole story is suspect. A fellow-biologist said to me: "You realise that it's the Peppered moth *larva* that is actually selected, not the adult?" If he was not joking, an awful lot of text-books are wrong.

4. From *Orthodoxy* 1908 by G K Chesterton

Chapter 2

The Evidence for Evolution

This question of evidence is a curious one. I saw a television programme recently in which a zoologist was walking about in a museum full of skeletons and telling us that they were evidence for evolution. I know what he meant: the bones are evidence in the sense that they are consistent with evolution. Another way of putting it is that evolution is a simple and elegant explanation of the similarities between the bones. But there is a difference between *fact* and *evidence*. It is a fact that the bones exist and another fact that they can be arranged by the curators in the order of the classification. The zoologist in the programme used the skeletons as evidence for Darwinian evolution. However a creationist would claim that they were evidence for Creation, and a Lamarckist would use them as evidence for Lamarckian evolution. Even if evolution is accepted, there is no evidence for an evolutionary *mechanism*: there is nothing about the skeletons to show that they evolved by a process of natural selection.

What is the evidence for evolution? Here is the evidence as we presented it to our pupils.

Evidence from Classification

Most objects can be classified, i.e. sorted into categories or groups. Collectors know all about this: many people collect coins or Staffordshire pottery. (When I was a schoolboy a number of us used to collect beetles. I didn't know it then, but we were following in Charles Darwin's footsteps; he was an enthusiastic "Beetle Man.") Once you have collected your specimens you have to put them in groups, to *classify* them; otherwise you cannot manage them, or even think about them in a logical way. An "artificial" classification is made just for convenience and usually doesn't tell you much about the things you have classified. Examples would be a stamp collection arranged just by colour (all the red stamps together) or a child's Wild-life book that put bats and dragonflies together because they have both got wings. A "natural" classification, on the other hand, is one that reflects deeper, more "real" similarities and differences. A sensibly arranged stamp collection would have the stamps arranged by country and date; a better wild-life book would put the bats with other mammals and dragonflies with other insects.

Most modern biological classifications are natural ones. They are also "phylogenetic" which means that they are put together to reflect underlying evolutionary relationships. The fact that this is possible is seen as good evidence for evolution.

Pupils found this rather unconvincing, partly because their knowledge of the Classification was scanty (all those Greek and Latin names). Perhaps the more logical students detected what looks like a circular argument: you invent a classification based on evolutionary relationships, and then use it as evidence for evolution! Of course, it's not really a circular argument. It is perfectly sensible to devise a tentative phylogenetic classification, fit the various organisms into it and then see how things work out from there. If new facts show that you have made a mistake, then - as a good scientist - you modify or abandon the classification. But the opposite has happened. Each new advance in knowledge, particularly recent work in genetics and DNA sequencing, has shown that this classification works rather well. There is no one-to-one relationship between an organism's characters and its DNA, and there has had to be some shuffling between the groups as knowledge has increased, but there have been few big surprises. Another point is that the natural classification consists mostly of "nested boxes": *all* rats are rodents, *all* rodents are mammals, *all* mammals are vertebrates, *all* vertebrates are animals; and these differences and similarities are echoed in the details of biochemistry and genetics, and this is what you'd expect if evolution is true.

We found that you had to be very careful of your wording when expressing these ideas as *evidence.*

Evidence from Comparative Anatomy and Physiology

Frogs and birds and horses (for example) have much the same bones in their skeleton; bones with the same names such as *thoracic vertebra* or *femur*, doing much the same jobs in the different animals. The muscles, blood vessels and nerves are arranged in similar patterns, work in the same way, and are given the same names in the various animal groups. The very cells of these animals are much the same, and have the same structures ('organelles') inside them, performing much the same biochemistry. Underlying everything is "the same" DNA. The most economical explanation of these similarities is that these plants and animals are somehow related. And the most obvious relationship is that they are joined by "blood ties", i.e. that they share a common ancestry.

Pupils found this to be the best evidence for evolution, and the enthusiasts among them would fill many pages of their notebooks with diagrams of insect mouthparts, the brains and blood systems of different vertebrates and the famous *pentadactyl limb*. (This "five-fingered" limb refers to the legs, arms and wings of the vertebrates. Of course not all these actually have five fingers, but there is a basic pattern and different organisms show various variations on the theme. We are so used to our own bodies that we forget how remarkably they are like the bodies of other vertebrate animals. Next time you are on a beach or at a swimming pool, you can have a diverting time looking at people's bare feet and marvelling at all those sets of five toes.)

Vestigial Features

A vestigial structure is one which is small, insignificant or apparently useless in one organism but which is large, or important, in a different organism, particularly one that might be an ancestor. The famous example is the human appendix. Many plant-eating mammals have a large, working appendix full of micro-organisms which digest the cellulose in the animal's food. The human appendix, however, is small and appears to have no function. There seems to be no reason for it to be there except as a vestige left over from the past, when our ancestors were herbivorous. My old 1940s textbook claims that there are 200 vestigial features in the human body. Goodness knows what they all are. After we had listed eyebrows, third eyelid, coccyx, goosebumps and some technical things like the ductus arteriosus, we began to run out of instances. There are plenty of other examples of vestigial structures in biology. An animal list would include vestigial wings in kiwis, vestigial pelvic girdles in both pythons and whales, vestigial eyes in some cave animals and the vestigial second pair of wings (halteres) in true flies. Plant examples include vestigial petals in grass flowers, and the scale-like leaves of some parasitic plants (like the dodder) and saprophytes (such as the bird's-nest orchid).

This is telling evidence for evolution, but it is hard to present these facts as evidence rather than as conclusion. We found that we had to be very careful with the wording. I have a text-book in front of me and although the heading is "Evidence for Evolution", all the descriptions of vestigial features *assume* evolution:

"this organ has *become modified* . ."
"the bird . . has *lost* the power of flight",
"these bones are *reduced to* splints."

Evidence from vestigial features particularly upsets creationists. The word "vestigial" implies evolution even before you start discussing it. They lean heavily on the point that many "so-called" vestigial structures actually have a *function*: they claim that if a structure has a perfectly good job to do it is wrong to call it vestigial. The coccyx ("tail skeleton" in man), for example, is important as an attachment site for some muscles of the pelvic floor. The pineal gland in the brain secretes hormones. Evolutionists reply that the function is neither here nor there; the important thing is that the coccyx *is* a tiny tail, equivalent to the longer tails of other vertebrates, and that it is best explained as a *remnant* of the tail; in the same way, the pineal gland is a remnant of the third eye that was a feature of some primitive vertebrates. These arguments have not proved fruitful and should probably be dropped.

Embryology (Developmental Biology)

Just as adults resemble each other in their structure and physiology, so do embryos. If you follow the development of different animals (fish, chicken, rabbit, human, say) they all start off in the same way. The fertilised egg divides to form two cells, then four and so on. Well of course that happens - how else could you begin to get from a single egg to a multicellular organism? But there are strange and apparently unnecessary similarities. The four different embryos look so alike at an early stage that it would take an expert to decide which animal any embryo was going to become. As development proceeds the embryo of a "higher" animal appears to go through the "lower" stages, looking at first fish-like, then reptilian, then mammalian, and so on, thus reflecting an evolutionary progression. Our textbooks were a bit cagey about this evidence because it was associated with a dubious idea called *Recapitulation*.

In the nineteenth century many embryologists believed that higher animals, during their development, actually went through the adult stages of their ancestors, that their embryological development repeated or "recapitulated" the detailed evolutionary steps that their ancestors had taken. So a human embryo went through an actual fish stage, a reptile stage, an early mammal stage, and so on, gradually becoming more human in the process. These stages or phases had somehow become condensed into the early development of the embryo. This idea is associated with Haeckel, a German professor of zoology and enthusiastic Darwinist. [1] There were several things wrong with this theory. Firstly, mammal embryos do not resemble *adult* fish or reptiles but only their embryos. Secondly, Haeckel and his supporters pushed their views too hard and set embryology off in the wrong direction for decades, with embryologists trying to find detailed one-for-one correspondences between, say, adult fish and embryo mammals, forgetting that a fish is adapted to one particular mode of life and a mammal embryo is adapted to a completely different mode of life within the womb. Thirdly, and worst of all, Haeckel was such an enthusiast for the idea of recapitulation that he illustrated it with some faked diagrams. His various forgeries were exposed, but he managed to get away with it by explaining that everyone fills in missing bits here and there and that most biological diagrams are "doctored" to some extent! (This was over 100 years ago, mind.)

So "Recapitulation" became almost a dirty word, and the textbooks tended to downplay the evidence from embryology, presumably to avoid any suspicion of contamination with Haeckel's ideas. This was a pity: embryology provides good evidence for evolution. Darwin (who did not himself accept recapitulation) certainly thought so. In a letter to Asa Gray, of September 10, 1860, he wrote:

"Embryology is to me by far the strongest single class of facts in favour of change of forms."

This is supported by new discoveries. Evolutionary developmental biology (evo devo for short: see Chapter 7) is a promising field which is breathing fresh air into biology.

Palaeontology

Palaeontology provides the only concrete evidence of what happened in the past. You can actually handle a fossil, examine it and study the rock in which it was found. Some scientists downplay the fossil evidence for evolution, but we would not know that ammonites, trilobites and dinosaurs had ever lived at all, if it wasn't for their fossils. In the biology department at Millfield we had a few nice fossil specimens and one or two good videos. Rather than let the students loose on thousands of disconnected facts we sorted the evidence into three main groupings:

1. The geological record

The record itself points to evolution: the fact that there are fossils at all, the fact that 95% of fossils are different from living forms, and the fact that they mostly appear in sequence. The vertebrate fossils tell a particularly clear story: there are no jawed fishes before the Ordovician period, no amphibia before the Upper Devonian period, no reptiles before the Carboniferous, and no mammals before the Upper Triassic.

2. Fossil series

Some groups of fossils can be arranged in a *series* of forms, A^1, A^2, A^3 . . , which show gradual changes from one type to another, usually in progressively younger rocks. In our teaching we mentioned several of these, such as the series of sea urchins and mammal-like reptiles, but we put most emphasis on the horse series. We had a good display presentation of these, as mentioned in Chapter 1, complete with casts of the lower leg bones of many of the genera, diagrams of the patterns of the enamel of the molar teeth, maps of the area where they were found, and so on. (We did admit the imperfections of the horse series: that the fossils were spread over different continents; that they didn't form a linear series but zig-zagged about; that they were not a true genealogy. We even showed our pupils a video from a *Horizon* programme entitled "Did Darwin get it wrong?" in which the presenter was rude about the over-simplification of the series as shown in some museum displays.)

3. Transitional organisms

These are organisms that bridge gaps between large groups. (Where the gaps are unfilled, various "missing links" are spoken of: we discussed these and why they might be missing.) We majored on the bird *Archaeopteryx*, which is so reptilian that if its feathers had not been preserved it would probably have been classified as a small dinosaur. The skeleton of *Archaeopteryx* was found in 1861, only two years after the publication of Darwin's *The Origin of Species*.

This fossil evidence was another of the evidences for evolution that our pupils really took to: some of them made many pages of detailed notes.

Although palaeontology provides the only hard evidence of creatures which lived in the past, it is controversial (see Chapter 12).

Geographical Distribution of Living Things

This is indirect evidence, like that from the Classification. It's as if you should say, "Let's *assume* evolution; now let's examine the distribution of animals and plants in the world and see if it all makes sense." Sure enough, it does. Geographical evidence is also of historical interest, because it is associated with Darwin and his long voyage in the *Beagle*. As with Classification, many pupils found this evidence unconvincing, but dutifully made notes on clines and Darwin's finches.

We had some good display material on the "Adaptive Radiation" story from Australia. For readers unfamiliar with this, it goes as follows: There is fossil evidence that marsupial mammals (pouched mammals such as kangaroos) appeared earlier than placental mammals (mice, cows, humans) and that they could not compete with the placentals in most of the world. Therefore they died out: but not in Australia, which became separated from the rest of Asia before the placentals appeared there. So just as placental mammals "radiated" to fill all the vacant niches (habitats and roles in the environment) in most of the world, the marsupials radiated in Australia free from competition with the placentals. This resulted in a series of marsupial animals which were counterparts to many of the placentals. The kangaroo is equivalent to a grazer like a deer; the Tasmanian wolf was a top carnivore like a wolf; hare wallabies are like rabbits; koalas are like small bears; and there are marsupial anteaters, mice and moles. The pupils were impressed with this story, but because they had not heard of many of the examples (wombats, bandicoots and such) the details did not always lodge in their memories.

The A Level geographers became specially animated when we talked about Continental Drift. They knew at least as much about it as I did because they had studied plate tectonics in geography.

This list of evidences may be seen now as old-fashioned. Since I retired from teaching there have been exciting new developments, especially in genome mapping and evo devo. If I was still teaching evolution I would present the evidence in a different way. But there are two reasons why I have left this chapter in its original form. Firstly, there might be some historical interest in the content of a rigorous A Level syllabus as it was in the second half of the twentieth century, before curriculum developers got their hands on it. Secondly, the arguments about evolution and creation in almost all the books available at the moment are also very old-fashioned, going over the same old ground over and over again. Many of the people I have talked to recently about evolution have never even *heard* of evo devo, facilitated variation or the other new ideas.

The importance of wording

It is important to use the right words when giving evidence for evolution. Words like "higher" or "advanced" are loaded and should not be used when *giving evidence*. You must not state that an organ is vestigial or "derived from" another organ, for you will be assuming the very thing you are trying to prove. Here is an insect without wings: did it *lose* its wings during the course of evolution or did it have none to start with? We must not call an organism or structure "primitive" if all we mean is "simple." "Primitive" means *earlier*; it implies a comparison with something that came later in time; "simple" just means *uncomplicated*.

Occasionally a pupil would write something like: " . . . so ducks evolved webbed feet for swimming" and we would discuss what was wrong with the wording. Firstly it sounds as if the ducks themselves had worked out what to do. Secondly, they weren't "ducks" at the start of the process. Thirdly, the English style is unfortunate: "to evolve" means "to change" and the birds didn't *change webbed feet*.

Another pitfall resulting from careless wording is that of personalising evolution, so that it becomes Evolution, with a capital E. We used to show a video about childbirth to our pupils. It was an excellent film, but it contained one extraordinary statement: "Most human skills have to be learnt, but evolution has seen to it that Amy is born with a few essential abilities . . . to give her a good start in life." Now there's an intriguing phrase: "Evolution has seen to it . . ." One pictures Evolution as a rather bossy goddess; a younger version of Mother Nature perhaps?

Microevolution

This list of evidences does not include the small-scale changes that many organisms can be seen to undergo. These may be called "evolution-in-action" or *microevolution*, for example the production of the different dog breeds, or the changes in wing-colour of the famous peppered moths that feature in all the textbooks. Instead we placed this small-scale evolution in the section "The Mechanism of Evolution" (see next chapter) because it is fast enough to see *how* it happens. However many authors put these small-scale changes in the list of evidence for evolution.

Comments on the Evidence for Evolution

Some scientists are so convinced of the truth of evolution that they tell us that we should not be concerned with these "pedagogic 'proofs'" of its past occurrence.[2] It is really a waste of time, they say, because biological thought and evolutionary thought are now the same thing. We should no more be bothering about the evidence for evolution, we are told, than about those "proofs" of the roundness of the earth which were once taught at school.

Not everyone would agree. There are plenty of people, including many highly-qualified scientists, who believe that the matter of evolution has not been settled. Schoolmasters have to start somewhere, and "Evidence for . . . " is not a bad introduction

Secondly, although this evidence is *for* evolution, it is not *against* other theories to account for the existence and variety of living things. It is not against Creation, for instance. Human beings have always known that our bodies are like those of other animals; that we need food, that we bleed when we are cut. Modern scientific research has added great detail, but philosophically not much has changed. We always knew we were made of *something*. There were classifications in place long before Darwin's time. Much of the evidence for evolution is actually evidence for the *relationships* between organisms, rather than for the explanation of the relationships. Creationists may not dispute the relationships but question the explanation.

Thirdly, some of the evidence for evolution seems excellent from a distance but begins to dissolve when you look at it in detail. Take comparative anatomy for instance. Parts of the body that are "the same" in different animals are called *homologous* (the radius bone in your forearm is homologous to the radius in a bird's wing). But when you study homologies in detail the argument begins to unravel. Homologous structures do not necessarily come from the same parts of the embryo, and may be governed by different genes. A human arm and leg are homologous with each other in the sense that they contain the "same" bones in the same pattern, but the arm is not evolved from the leg. The argument then becomes complicated, and sooner or later someone will say (I've seen it in print) that the evidence from homology is not very important [3]. Again, the evidence from fossils is the only actual concrete evidence we have from the past, but in many respects it is non-Darwinian (see Chapter 12). Several authors are beginning to say that we need not rely on fossils because there is much better evidence from elsewhere. [4] To Darwin and Haeckel, the evidence from embryology was the most compelling of all, but it is now downplayed and discrepancies have to be explained away. I am quite sure that if you found problems with, say, the geographical evidence for evolution, someone would tell you that it didn't matter because there was much better evidence elsewhere. It reminds me of Alice in the shop in *Through the Looking-Glass*: " . . whenever she looked hard at any shelf, to make out exactly what it had on it, that particular shelf was always empty: though the others round it were crowded as full as they could hold." It will be fascinating to see if the same thing happens with the evidence that is now being adduced from gene sequencing.

Evolution is a "Big Idea." It isn't that there are one or two very strong arguments for it, but that there are numerous different arguments from different branches of biology. The Lilliputians held Gulliver down not by one or two strong cables but by hundreds of tiny threads. This multiplicity of little evidences, from many branches of science, may be called "consilience" (see Chapter 15). This explains something that might puzzle opponents of evolution: the fact that no one takes very seriously the odd apparent counter-example. I have seen pamphlets claiming that the highly-complicated spinnerets of a spider, or pieces of gold chain found in a coal-seam, *disprove* evolution. To someone who accepts evolution these things don't signify. If Gulliver is held down with a million threads, breaking three won't release him. This is not a matter of taking sides, it's just common sense.

Notes to Chapter 2

1. Haeckel coined the word *ontogeny* to mean the development of the individual organism (= its embryology) and *phylogeny* to mean the history of the race (= its evolution) His idea of Recapitulation became shortened to the slogan "ontogeny recapitulates phylogeny" (also called the "Biogenetic Law.")

2. *The Life Science* (Wildwood House, 1977) by Peter and Jean Medawar, Chapter 2.

3. For details see *Homology, an Unsolved Problem* Oxford Reader, 1971 by Gavin de Beer.

4. Mark Ridley wrote: "No real evolutionist . . uses the fossil record as evidence in favour of the theory of evolution as opposed to special creation." and later, "the gradual change of fossil species has *never* been part of the evidence for evolution." *New Scientist*, June 1981.

Chapter 3

Mechanisms of Evolution

Lamarckism

Having spent some time teaching the evidence for evolution, we went on to discuss the different mechanisms that have been proposed to explain evolution, starting with that of Lamarck.

In 1809 (the year that Darwin was born) Jean Baptiste de Lamarck published his *Philosophie Zoologique* which included an idea called "The Inheritance of Acquired Characters." He pointed out that animals changed during their lives (e.g. some muscles get stronger) and he suggested that such changes could be passed on to their offspring. Hence a group of animals could become better fitted to their environment over time. The famous example, always associated with Lamarck, was that of the giraffe's neck: by continually stretching its neck to get at the higher leaves on trees and bushes that were out of the reach of its competitors, the giraffe would gradually lengthen its neck during its lifetime. This increase in length would be passed on to the next generation (it was supposed) and so, over thousands of years, giraffes ended up with very long necks.

This scheme turned out to be wrong. Most characters of an animal cannot be changed by "use or disuse." A weightlifter can get bigger muscles by exercising, true, but how could a zebra *acquire* camouflage or an invertebrate *acquire* a backbone? Secondly, the very few characters that can be acquired (such as nimble fingers in a pianist) are simply not inherited. It is impossible not to feel sorry for Lamarck. He was badly treated by his contemporaries and has been mocked by posterity, but actually he was a very good zoologist, and the theory of the inheritance of acquired characters was only a small part of his work. He cannot be blamed for his mistaken views on inheritance, because all his fellow-biologists were in the same boat: Mendel's work on genetics was not discovered until nearly a century later. And the idea that acquired characters might be inherited is not a stupid idea; it was widely believed in the nineteenth century and Darwin accepted it in part.

It has been suggested that teachers should not waste time telling students about Lamarckism because it is known to be wrong. But we found it useful to consider *why* it is wrong, and to link it to discussing how DNA works

It is strange how the standard version of Lamarckism always involves giraffes. If you read the text-books you might think that the world was faced with this grave problem: how do you explain the giraffe's long neck? First there's the Lamarckian story, now shown to be all wrong, followed by Darwin's true explanation based on natural selection.

But the whole giraffe story is extremely dubious. Lamarck only mentioned the giraffe in passing; he had much better examples of his theory. In the early editions of *The Origin of* Species Darwin does not mention the giraffe's long neck at all; in the sixth edition (the one we have today) he does refer to it but his explanation is a combination of Lamarckian inheritance and natural selection. [1]

There are other problems. Firstly, if the giraffe's long neck is essential for survival, what about female giraffes (much smaller than the males) and all the foals? Secondly, this is all guesswork. There is no fossil evidence to show how their necks evolved. Thirdly, the whole thing is absurd: you cannot *stretch* your neck! Muscles can only contract, they cannot push. (Put your hand down on a flat surface and try to stretch your middle finger and see what happens.)

It makes much better sense to suppose that the long neck is a consequence of having long legs. Giraffes have long legs for fast running (arrived at by whatever mechanism you like) and they need to get their mouths down to the water in order to drink. Not having a trunk, like the comparably tall elephant, they need a long neck.

The idea of the Inheritance of Acquired Characters has a strange persistence. There is something obvious, even elegant, about it. And wouldn't it be nice if it were true? If a child could actually inherit not just his parent's aptitude but the actual skills produced by years of practice? It is an idea which has not quite died, and every now and then you hear of someone, somewhere, doing a "Lamarckian" experiment. A modern biologist would say that the idea is absurd. The whole of genetics and the biochemistry of DNA are against it. But are they? What about nonrandom mutations induced by the environment? What about cytoplasmic inheritance (inheritance which bypasses the nuclear DNA)? A fertilized egg has a large volume of maternal cytoplasm and it contains a number of self-replicating bodies, such as mitochondria (mitochondria and chloroplasts have their own DNA). If some acquired character affected the organism's mitochondria, it could be passed to the next generation through the cytoplasm. (A biology lecturer once remarked to me, "Cytoplasmic inheritance is the thin end of the Lamarckian wedge.") It has been suggested that the whole cell is the unit of inheritance, rather than the gene. This - if true - would put all these theories back into the melting pot. [2] We may not have heard the last of Lamarck.

Natural Selection

We taught this as Darwin's great contribution, but it is worth remembering that several other people had come up with the same idea, notably the biologist Alfred Russell Wallace (see Chapter 15).

In the first chapter of *The Origin of Species* Darwin wrote about the breeding of domestic pigeons, in order to introduce the idea of selection. The breeder selects as parents of the next generation those birds possessing the qualities he wants his stock to have, and at the same time he prevents the reproduction of the ones that lack the desired qualities. This artificial selection has therefore two aspects, one positive and one negative. Darwin claimed that nature exercises selection in much the same way, picking out winners and losers in the race to become parents of future generations.

The usual way to teach natural selection is to make a list of facts, with deductions from them, along these lines:

Fact: Organisms overproduce (i.e. there are normally far more offspring than parents). However, populations normally stay much the same size (so for each pair of parents there will just be two ultimate survivors).

Deduction: *There is a high mortality rate (i.e. there is a struggle for survival).*

Fact: Individuals in a population vary in their characteristics.

Deduction: *Some will be better adapted to their environment and will therefore be more successful in the struggle for survival.*

Fact: Many of these variations in characteristics are heritable.

Deduction: *The good variations will be inherited; therefore the population as a whole will become better adapted; this change is evolution.*

What could be more sensible? No wonder that Thomas Huxley, Darwin's friend, is said to have remarked: "How stupid not to have thought of that oneself!" In fact, natural selection is so sensible that many people have accused it of being just a tautology, especially when it is described by the catchphrase "The Survival of the Fittest." This formulation of natural selection can be seen as a circular argument along these lines:

Question:"Which organisms survive?"
Answer: "The fittest."
Question: "How do you know they are the fittest?"
Answer: "Because they have survived."

I expect you can see how this word-trick works.[3]

Natural selection is just common sense and universally accepted. Even creationists agree that natural selection happens, and that it gives rise to different forms of wild animals or plants in the same way that artificial selection produces different breeds of dogs or varieties of roses. However some authors doubt the scope of natural selection because they question the truth of three of its supporting statements:

(a) Organisms do not always "overproduce." Populations are not necessarily limited by external mishaps (predation, starvation, disease and so on) but may be controlled by innate adjustments to breeding numbers. For example, plants which are crowded together make fewer seeds.

(b) Competition is not universal; there are numerous examples of cooperation and many species coexist peacefully.[4]

(c) Natural selection does not result in continuous, endless change but soon hits natural limits (you cannot breed a rat the size of a sheep).

As a teacher I discovered an odd fact about natural selection: here is this universal moulder of living things, this proposed cause of all the variety in the living world around us - and *it is very hard to find examples*. In the biology department, we had a wide collection of teaching aids (charts, specimens, videos &c.) on the other aspects of evolution; but we found very little on natural selection. We tried to get away from the text-books, but were unsuccessful and ended up having to use the same handful of examples, namely:

The peppered moth, *Biston*, with its speckled and dark (melanic) forms.

The snail, *Cepaea*, with its different colours and patterns of banding.

DDT-resistance in Diptera such as mosquitoes.

Antibiotic-resistance in bacteria.

Heavy metal-tolerant grasses growing on spoil heaps.

And that's about it. There are other examples in nature, but they are mostly little-known organisms studied by specialists. They cannot be very convincing or they would have found their way into the textbooks. Why is it so hard to find examples?

The above list is surprisingly short and it has other odd features. All but one of the above text-book examples of selection are from the result of human interference with nature. It is curious that *natural* selection should be best demonstrated from *unnatural* situations. Secondly they are not very good examples. This year I have seen countless thousands of *Cepaea* snails (but, interestingly, not a single thrush, once their main predator) and they were present in all their different colour forms and different banding patterns. They do not seem to have actually *evolved* at all. As for changes in bacteria, such as antibiotic-resistance, Professor Grassé has pointed out that bacteria are not very good examples of evolution, because they stabilised a billion years ago and have scarcely changed since!

Here is a condensed version of the peppered-moth story:

The peppered moth (*Biston betularia*) is a British moth with a wing-span of about 4 cm, the larvae of which feed on birch and lime. The wings of the normal form are greyish-white, speckled with black but there is also a 'melanic' form with almost black wings. Peppered moths may rest by day on surfaces such as tree trunks. In open country, where lichens grow on the bark the pale moths are well camouflaged, but in industrial areas, where pollution has killed the lichen and the bark is sooty, it is the dark, melanic moths that are hard to see. Observation by naturalists revealed certain facts about the presence and distribution of the moths, and these have been confirmed by experiment. Firstly, the less well-camouflaged moths are more likely to be eaten by birds. Secondly, in rural areas it is mostly the pale form that survives. And thirdly, as pollution increased in industrial areas more and more melanics survived and the pale ones were eaten. This change was dramatic: in 1849 the melanics formed just 1% of the population; in 1900 they formed 98% of the population near cities like Birmingham and Manchester. Note that eighteenth century moth collections have the occasional black peppered moth in amongst the normal pale forms. This shows that the melanic mutation is not a direct *response* to the newly polluted environment.

The peppered moth story has been hailed as "evolution in action" but, strictly speaking, it is just natural selection in action. The peppered moth has not actually *evolved*: since the various Clean Air Acts have been in force, the numbers of the pale form have increased again. Instead, peppered moths show polymorphism [5]: they naturally have pale and dark forms which fluctuate: where the environment is sooty the dark ones predominate and where the environment is clean the light ones predominate.

These small-scale changes come under the heading of *microevolution*. Large-scale evolution (reptiles giving rise to birds, say) is called *macroevolution*. Originally, *microevolution* referred to the evolution of subspecies and geographic races and *macroevolution* referred to the evolution of species, genera, and all higher groupings. Nowadays people seem to use microevolution to refer to any small-scale change. The radiation of Darwin's finches is referred to as microevolution, even though they are separated into five genera and fifteen species. There is no agreement about where to draw the line between micro- and macroevolution. Here is a mischievous definition to argue about:

Microevolution: any evolutionary change shown to be caused by natural selection

No one doubts that microevolution occurs, and creationists accept it readily.

The neo-Darwinian Synthesis

The neo-Darwinian Synthesis (NDS) is the modern formulation of evolution: Darwin plus genetics. Darwin did not know the source of the variation on which natural selection acted. In his day there was no science of genetics: inheritance was not understood. Mendel had published his work on hybridisation in pea plants during Darwin's life-time, but it was in an obscure journal, and no one took much notice of it until after Darwin's death. (Darwin died in 1882, and Mendel's work was not discovered until the turn of the century.) For seventy years after the publication of *The Origin of Species* scientists argued about Darwin's ideas. When the new science of genetics was established in the first decades of the twentieth century many scientists moved away from Darwin. New knowledge about chromosomes and mutations made his views look old-fashioned, and some of the pioneers of the new genetics thought their work had proved Darwin wrong. By the 1930s his theory was in disarray. Then a group of scientists working together revised Darwin's ideas in the light of the new knowledge. [6] The result was *the Modern Synthetic Theory of Evolution*, known today as the neo-Darwinian Synthesis (NDS). This allied Darwin's theory with the new genetic discoveries: the immediate source of variation is the gene shuffling associated with gamete formation and fusion during sexual reproduction. The ultimate source of variation is mutation, now known to be changes in DNA.

By the time that we taught evolution to our A level pupils they had already studied both genetics and the biology of DNA in detail, so these ideas could be readily linked.

The textbook line is that (i) there is plenty of evidence for evolution , (ii) there is plenty of evidence for natural selection and (iii) *the former is caused by the latter*. It was not until I had almost come to the end of my teaching career that I realised that statement (iii) is an assumption supported by argument rather than evidence.

Notes to Chapter 3

1. See Stephen Jay Gould: 1998 *Leonardo's Mountain of Clams and the Diet of Worms* Jonathan Cape; Chapter 16 "The Tallest Tale."

2. For recent developments in "epigenetic inheritance" (inherited changes that by-pass the DNA) see the article "Strange Inheritance" by Emma Young, (*New Scientist*, 12 July 2008)

3. If not try this similar set of questions and answers:
Question:"Which candidates passed the exam?"
Answer: "The best ones."
Question:"How do you know they are the best?"
Answer:"Because they passed the exam."
Plenty of things can be ridiculed by writing them in an apparently circular form. The moment you add details (the best candidates are those with the greatest knowledge, and thus the highest marks) the statements make sense.
Here's something that used to amuse us from our student days:
Question: "Why do undergraduates have to wear a gown after dark?"
Answer: "So that the Authorities can see whether they are members of the University or not."
Question: "But why do the Authorities want to know whether they are members of the University?"
Answer: "So they can make sure that they are wearing their gowns."

4. Some Darwinists argue that co-operation is the *result* of natural selection. This might reflect reality, or might be special pleading. How can you tell?

5. *Polymorphism* is the existence of adults in more than one form: milkwort (*Polygala*) flowers are blue, pink or white; *Cepaea* shells can be brown, pink, yellow or banded; peas may be tall or short, and so on.

6. The Synthetic Theory was established in the 1940s by Stebbins and Dobzhansky (geneticists), Fisher and Sewall Wright (mathematical geneticists), Mayr and Julian Huxley (zoologists) and Jepsen and Gaylord Simpson (palaeontologists). A key date is 1942 which marks the publication of Julian Huxley's influential book *Evolution: the Modern Synthesis.*

Chapter 4

Macroevolution and the neo-Darwinian Synthesis

There are conflicting ideas about how macroevolution happened. The story in the books is that it is simply an extrapolation from microevolution: they have the same mechanism. Well, why not? Unfortunately, although there are plenty of arguments for this claim, there is almost no empirical evidence. Neo-Darwinists assert, with some impatience, that *of course* it's the same process. They will refer you to Occam's Razor, to the multiplication of hypotheses. They will say that natural selection is far too elegant an idea for it *not* to be the cause of all evolution. They point out that the reason there's no evidence is that evolution takes so long, and that a *process* cannot be fossilised. Other evolutionists are less sure. Yes, evolution has taken place but does the orthodox mechanism explain it? Can natural selection do more than make minor adjustments to what is already there?

In the 1980s and '90s some books were published (not by creationists) that questioned the neo-Darwinian synthesis.

The Neck of the Giraffe (subtitled *Where Darwin Went Wrong*) by Francis Hitching appeared in our department library in the early 1980s. Th author's aim was to expose cracks in the Darwinian explanation. Hitching discussed the non-gradual nature of the fossil record, the undue emphasis on genetics rather than how an animal or plant is actually built and the way evolution has become a holy cow amongst certain biologists. He agreed with those scientists who think that you cannot simply extrapolate from microevolution to macroevolution. I read the book with great interest and I think that some of my colleagues did as well, but it made little difference to our teaching.

The Intelligent Universe by Fred Hoyle, also appeared in our department library. You could get an idea of what the book was like from the jacket blurb, which asked whether life started by random processes and whether Darwin's theory of evolution was still plausible and answered them with a clear "No." But a quick browse through the book shows that Hoyle was presenting an unorthodox theory of his own. Chapter headings like "The Interstellar Connection" and "Evolution by Cosmic Control" reflected ideas that were too eccentric for school biology teaching. My colleagues never mentioned the book, and I doubt if most opened it. But Hoyle's criticisms of Darwinian evolution are worth looking at. In a chapter headed "The Gospel According to Darwin", Hoyle reviewed the evidence for evolution itself, and attacked the current theory, concluding that Darwinian theory is clearly wrong because random variations tend to worsen performance.

A third book was *Evolution - A Theory in Crisis* by Michael Denton, a molecular biologist. This book, full of scientific data, is a critique of neo-Darwinist theory, showing that it is not supported by the evidence.

From these books I discovered that there were many scientists (Hoyle, Grassé, Stanley, Saunders and Ho, Gould and others) who simply did not accept the NDS in its current form. These books were not written by creationists. The authors did not attack the idea of evolution itself, but argued that the proposed *mechanism* was inadequate: that the NDS could not explain macroevolution. You could summarise their position as: "Evolution? Yes; neo-Darwinism? No." I was familiar with EPM literature that attacked evolution, and knew of creationist activity in America. But here were scientists attacking not evolution itself, but the neo-Darwinist explanation of evolution, and without any reference to the age of the earth or the Creation account in Genesis.

These books, by Hitching, Denton and others were not "Attacks on Darwin" but attacks on what others have done with Darwin's ideas. If you go back to what Darwin actually wrote in *The Origin of Species*, it looks as if he got it about right and that it is his disciples who are getting it wrong. Darwin was cautious in what he wrote, and honest about counter-arguments. He insisted that natural selection was not the only mechanism in evolution [1] and when he wrote of variation being "random" he meant that its cause was unknown:

"I have hitherto sometimes spoken as if the variations - so common and multiform with organic beings under domestication, and in a lesser degree with those under nature - were due to chance. This, of course, is a wholly incorrect expression, but it serves to acknowledge plainly our ignorance of the cause of each particular variation." [2]

Nevertheless, there are two ways in which some biologists consider that Darwin himself *did* get it wrong.

Firstly, he maintained that evolutionary change must be gradual. Even his greatest contemporary supporter, T H Huxley, thought that this was a mistake. When Darwin insisted that evolution must be gradual, that there must be no saltations ("jumps") in his scheme, Huxley told him "you have loaded yourself with an unnecessary difficulty." Supporters of the NDS maintain that Darwin was right, of course, and they have devised ingenious - and unconvincing - explanations for the gaps both in the fossil record, and between the phyla.

Secondly, Darwin originated a particular method of arguing in the *absence* of evidence. Natural selection is a simple idea, and it is as though its very plausibility removes the need for any proof. By mixing masses of detailed observation with clever speculation Darwin managed to convey a degree of certainty about his conclusions. Not everyone found this convincing. In his Introduction to the 'Everyman' *Origin of Species*, published in 1956, Professor W R Thompson wrote:

"The deficiencies of the data were patched up with hypotheses, and the reader is left with the feeling that if the data do not support the theory they really ought to." [3]

I would not like the reader to think that our department library was filling up with books attacking the NDS! Of the ones I have just mentioned, only those by Hitching and Professor Hoyle appeared on the shelves. Mainline evolution books by a variety of authors also came into the library from time to time, ranging from the dubious-but-entertaining (*The Naked Ape* by Desmond Morris) to the worthy-but-dull (*The Theory of Evolution* by John Maynard Smith). The biology staff were mostly too busy to read all this material, but we used the big, illustrated books as part of our display.

These books were circulating when I was still teaching biology, some years ago. A newer "attack on Darwin" - and from-self-confessed atheists (which makes it of particular interest) - is the recent book by Jerry Fodor and Massimo Piatelli-Palmarini in which they show how poor was the analogy that Darwin made between selective breeding and natural selection. "Selection" in the case of breeding is a *mental* activity by a human agent with a particular end in view. It is *intensional*. Nothing at all like this happens in nature, and therefore Darwin failed to elucidate the mechanism by which evolution takes place. [4]

Problematic features of the NDS

Evolution is gradual.

Darwinism requires continuity throughout the living world. But living things are discontinuous, with big gaps between groups. These gaps are not just in the fossil record, but between living forms, both in their structure and their biochemistry. In the fossil record organisms do not gradually change from one form to another. They appear, stay much the same for millions of years, then disappear. This feature of the record is so striking, and so disappointing to Darwinian expectations, that two distinguished American palaeontologists, Gould and

Eldredge, proposed a theory (called "Punctuated Equilibrium") to try to account for it.

It can also be argued that evolution cannot be gradual because for each change in a structure you have to have many concomitant changes elsewhere in the organism. A new hormone would need a new receptor, and so on. This means that any novelty would have to appear as a complete "package." A single feature could change gradually, but if many have to change at once, is not that a jump? This criticism is now being overtaken by new research (see Chapter 7).

The ultimate source of variation is mutation.

This is an assumption. It has not been established that mutation is the cause of all variation, only that it is the cause of some, present-day, variation.

Mutation is random.

"Random" may mean simply that the processes involved are not yet fully understood. Opponents of the NDS have also made these claims:

1. Copying mistakes (such as point mutations) are probably random but of little importance in evolution because even if they confer some small advantage they represent a loss of information from the genome.

2. Complex chromosomal changes (deletions, insertions, inversions &c) are not random because they can be precisely reversed by the cell when appropriate. [5]

3. There is evidence for *directional* mutations induced by the environment. [6]

4. When you actually do the maths, the claim that random mutations plus millions of years are enough to give the

required variation for cumulative natural selection to work on cannot be substantiated. Hitching and Denton discuss this at length. [7,8] Augros and Stanciu point out that there is no correlation between the accumulation of random mutations and the rate of evolution. During the same time that mammals evolved into 16 separate orders (producing animals as different as bats and whales) the thousands of frog species stayed frogs, forming just one order. Yet point mutations accumulate at the same rate in mammals and frogs. [9] Spetner devotes five chapters of his book to the question of randomness, claiming that the NDS is unworkable. [10]

These arguments have nothing to do with creationism nor the concept of Intelligent Design. They are *scientific* questions, to be decided in the laboratory.

Natural selection is the main agent.

Natural selection results from competition between organisms. Darwin was convinced of this, especially after reading Malthus's essay on human population, and he wrote at length about the struggle for existence. The very last paragraph of *The Origin of Species* speaks of a "Struggle for Life", of the war of nature, of famine and death. Television wild-life programmes often major on struggle and competition. But this emphasis may be misplaced. Augros and Stanciu have a whole chapter entitled "Cooperation" in which they show, with numerous examples cited from the biological literature, that many organisms are *not* struggling for survival. [11] Even where there is a struggle it is not clear that being well-adapted is the key to survival. Ladybirds don't pick out the weakest aphids, but the nearest. Plants highly adapted against fungus disease may be eaten by locusts. Is not most death or survival a matter of accident? Perhaps this is one reason why we found it so very difficult to get hold of good examples of natural selection for use in teaching.

Perhaps the difficulty of finding good examples of natural selection is because it is uncommon. If an oak tree drops 1000 acorns and only one grows up to become a young oak tree, is it because it is a genetic variant selected for its unusual "fitness" ? Or is it just that it happened to land in a suitable patch of soil where it can germinate, and that it wasn't eaten by squirrels and birds like the rest of the acorns? [12]

Neo-Darwinists insist that natural selection is a creative force. It does not just eliminate the unfit, but builds the fit, step by step. Other biologists maintain that natural selection is just a negative force, weeding out the unfit rather than creating novelty. Phillip Johnson wrote:

"Natural selection is a force for building adaptive complexity only when it is formulated as a tautology or as a logical deduction unconnected to any empirically verifiable reality. Whenever natural selection is actually observed in operation, it permits variation only within boundaries and operates as effectively to preserve the constraining boundaries as it does to permit the limited variation. The hypothesis that natural selection has the degree of creative power required by Darwinist theory remains unsupported by empirical evidence." [13]

Selection is an abstract noun, one we apply retrospectively. "These animals must be products of selection because here they are, alive in front of us." But can an abstract noun be an *agent*? The NDS treats organisms as if they were passive; at the mercy of their genes and their environment. "This finch evolved a strong beak" apparently means "A strong beak happened to this finch."

If it really is a creative force the term "natural selection" is unfortunate. Apart from the fact that only a mind can select, selection implies a pre-existing something to be chosen, and whatever produced that 'something' must have been the creative agent, not the subsequent selection. To put it another way, the dog breeder may have *selected* the colour of the red setter, but he did not create the colour. The characteristics chosen by any selective process are the products of a living system, an organism or a cell. They are not created by the selection process itself. Natural selection explains the frequency, but not the existence, of different types of animals and plants. We must look elsewhere for the creative agent.

Macroevolution is an extrapolation from microevolution; no other mechanism is needed.

Extrapolation is a dubious exercise because you go from the known to the unknown. (Having spent years teaching pupils how to draw graphs I am wary of extrapolation.) You don't know what limits there might be. Plant and animal breeders know all about limits in artificial selection (you cannot get a blue daffodil). Are there none in natural selection?

Other Criticisms of the NDS

Limiting Factors and characters "on hold"

We had in the biology department a video about the human excretory system. The presenter discussed practical things like drinking, sweating and excretion. At one point there was a (discreet) shot of a silhouette of the urine stream of a man relieving himself, and the narrator pointed out that the male urethra had become 'rifled', so that the stream is coherent and doesn't splash - a nice refinement of evolutionary design.

Here's the problem: the urine stream is indeed as described, but how exactly is it the result of natural selection? What contribution to "fitness" does it make? Even if it helped survival in some way, what about the poor females? Worst of all, is everything else "on hold" while this particular characteristic is being selected for?

This applies to all of an organism's characteristics. Is the evolution of the giraffe's digestion "on hold" while its neck gets longer? At any one time there must be some characteristic that acts as a limiting factor - the one that matters for survival *at that moment*. There is no point in a caterpillar being beautifully camouflaged if it is feeding on a poisonous plant which will kill it. A Galapagos finch might have a perfect beak but a rotten digestion. It is hard to believe that the rifling of the urethra was ever limiting; and if not, how could it be the product of Darwinian evolution? Professor Dover wrote: "The idea that what is important is a gene's average effect on the reproductive successes of its bearers, generation after generation, falls into the difficulty that evolution could not proceed by taking each gene, one at a time, whilst holding constant the effects of all the other supposedly selfish 99,999 genes which make up a typical 100,000 genes' worth of contribution to an advanced primate phenotype. Were this to be done for each gene in turn, then there would not be enough time, since the cooling of the earth, to have evolved what we know has evolved." [14]

Matters of life and death

The famous examples of natural selection are *life or death* matters: organisms survive or not, depending on their camouflage (peppered moths, *Cepaea* snails), beak shape (Galapagos finches) or resistance to chemicals (mosquitoes, bacteria). No one doubts that natural selection plays a part in those instances. But what about the hundreds of minor characteristics of an organism? In what way has natural selection caused your little finger to be shorter than your middle finger? The size of a finger does not affect survival, or the chances of becoming a parent. The concept of natural selection, so clear in the case of bacterial resistance to antibiotics, has been unwarrantably extended to cover everything about an organism. There is no empirical evidence for this because the past has gone and present-day experiments would be far too slow. A recent book referred to natural selection "working as a sweeper-up of countless little details." In the maelstrom of life and death and reproduction surely the "little details" would be completely invisible to the clumsy tool that is natural selection.

These criticisms of the NDS might be dismissed by Darwinists as being just the sort of thing that an ignorant opponent might come up with in a desperate attempt to unseat Darwinism. But those who have raised these problems are not ignorant. Real mathematicians have questioned the probability involved in the claim that mutation is random. Working palaeontologists have described the gaps in the fossil record. Research chemists have remarked that the spontaneous origin of life is a chemical nonsense. Computer experts have questioned whether randomness can generate information. The underlying paradigm has been criticised by professional philosophers. It was a Darwinist, Stephen Jay Gould, not a creationist, who once remarked that the synthetic theory (the NDS) was "effectively dead." [15]

A list of Darwinian facts and puzzles

1. There is much greater variation in DNA in the genomes of the 3000-odd species of frogs (all clearly frogs) than between that of the bat and the blue whale.

2. There is far greater variation in "domestic productions" as Darwin called them (dog breeds, fancy pigeons, &c.) than you get in the wild.

3. The gaps in both the fossil record and the classification are *in the wrong places*. You would expect few transitional forms between closely related organisms and many between distantly related organisms - but it's the other way round.

4. Darwin knew about that puzzle and he also knew about this one: most of the characteristics that are important in

the Classification don't seem to have much to do with survival or selection. It is hard to see how the number of petals in a flower, or bristles on an insect's leg have a part to play in determining "fitness."

5. If evolution is caused by natural selection, and natural selection is the result of competition, you would expect the fastest evolution when the competition is greatest. But it looks as if evolution is fastest when the pressure is *off*, for example after a mass extinction, when you have half-empty ecosystems with many niches available.

Finally, much is made of the fact that human beings share over 98% of our DNA with chimpanzees. Well, why not? Our bodies are made of organic chemicals and our cells have organelles common to most living things. The very respiratory pathways in our mitochondria are exactly the same as those of a cabbage. *Of course* we share a high percentage of our DNA with other organisms, particularly primates. It would make no philosophical difference if the figure was 100%: being human is not a matter of chemistry. So why is this seen as so important? If anything, it shows that similarities in DNA are not as significant as they are made out to be.

I do not wish to be misunderstood here. The discovery of the structure of DNA is the most important event in biology during my life-time. I enjoyed teaching the biochemistry of DNA. At Millfield we treated it partly historically, partly as a detective story, partly as an example of how science works. We showed pupils the *Life Story* video, and our copy of the 1953 Watson and Crick paper from *Nature*. We made models of, and devised practicals on, DNA. But that's it. The near worship of DNA in some quarters is plain silly. The DNA molecule is held up for our admiration as a *replicator*. But it cannot *self*-replicate, any more than a sheet of instructions can make a copy of itself without some form of copying machine. It is not alive; it is just a molecule. DNA is important, but it is not an agent; the *cell* is the agent.

This question of our kinship with chimpanzees ties in with a characteristic of Darwinist authors: their gleeful belittling of mankind. We are just a species of ape, we are told. Our planet is a minor world ("speck" is the usual term of dismissal) orbiting an insignificant star in an unremarkable galaxy; one of billions of galaxies. Marx has shown us that our aspirations are just a matter of economics. Freud told us that our noblest traits are just manifestations of sex and selfishness. Darwin toppled us off our pedestal as "Nature's last Word", and his present-day disciples tell us that we are no more evolved than the bacteria and that our very thoughts are a product of physics and chemistry. But even if we believe all that (and I doubt that many people do), what difference does it make? What is the point of saying all those things, unless it is the dubious pleasure of knowingness, a sort of playground one-upmanship?

Man's insignificance is not a modern discovery. Thousands of years ago a wise man wrote: "When I consider your heavens, the work of your fingers, the moon and the stars, which you have set in place; what is man that you are mindful of him . . . ?" [16]

This belittling of man, especially in relation to the stars, reminds me of a sentence or two from a book read in childhood. Dick, a serious young boy with a love for science, is looking up at the night sky:

"Those little stars . . were farther away than he could make himself think . . those tiny stars must be enormous. The whole earth must be a tiny pebble in comparison. . . He felt for a moment less than nothing, and then, suddenly, size did not seem to matter. Distant and huge the stars might be, but he, standing here with chattering teeth on the dark hillside, could see them and name them and even foretell what next they were going to do. "The January Sky." And there they were, Taurus, Aldebaran, the Pleiades, obedient as slaves . . .He felt an odd wish to shout at them in triumph, but remembered in time that this would not be scientific."

Winter Holiday Arthur Ransome

What a lot of wisdom packed into just a few words!

There are two fault lines in the Origins debate rather than one. Most people are aware of the two opposing positions of Evolutionism and Creationism which are discussed in Part 2. But there is a division within evolutionist ranks too, between those who adhere rigidly to the neo-Darwinian synthesis and those who think it is flawed. Many of the questions raised may be settled within the next few years as fresh knowledge about embryology, biochemistry and genetics comes in. [17]

Notes on Chapter 4

1. In the 6th Edition of *The Origin of Species* Darwin wrote:
"I am convinced that natural selection has been the main but not the exclusive means of modification." Darwin saw this as an important disclaimer, and so do others: the passage containing this sentence is referred to twice in *Alas Poor Darwin* (see Bibliography); once by Mary Midgley and once by Stephen Jay Gould.

2. *The Origin of Species*, Chapter V (opening words)

3. Thompson was critical of parts of Darwin's work. "To establish the continuity required by theory, historical arguments are invoked, even though historical evidence is lacking. Thus are engendered those fragile towers of hypotheses based on hypotheses, where fact and fiction intermingle in an inextricable confusion." (from The Introduction to *The Origin of Species* by Professor W R Thompson FRS)

4. *What Darwin got Wrong* (Fodor & Piatelli-Palmarini)

5. *Not by Chance!* (Spetner) Chapter 4 gives details.

6. Ibid. Chapter 7 This is not Lamarckism. Lamarck's idea was that the actual characters of the organism (*phenotype*) became altered by "use or disuse" and then inherited. Non-random mutation is when the *genotype* (the DNA itself) is changed by cellular processes which have responded to environmental cues. Spetner gives several examples.

7. *The Neck of the Giraffe* (Hitching) Chapter 3

8. *Evolution - A Theory in Crisis* (Denton) Chapter 14

9. *The New Biology* (Augros and Stanciu) Chapter 6

10. *Not by Chance!* (Spetner). Chapters 3 to 7

11. *The New Biology*, Chapter 4

12. Compare this with Jesus' parable of the Sower spreading seed. The successful seed was not *selected for* because it was the fittest; it happened to land on good soil.

13. *Darwin on Trial*, Phillip Johnson (Chapter 7)

14. from "Anti-Dawkins", the 4th Chapter of *Alas Poor Darwin*, (see Bibliography)

15. Stephen Jay Gould: "Is a new and general theory of evolution emerging?", *Palaeobiology*, 6, 1 (1980)

16. Psalm 8; verses 3,4

17. There is a faultline within Darwinism itself, between strict adaptationists and the more relaxed "Gouldians": see *The Darwin Wars* (1999) by Andrew Brown

Chapter 5

From Biology to Everything

During my time as a biology student and teacher, there has been a change in the way that many people think about evolution. This has been outside the classroom, I am glad to say, because it has been for the worse. The change is from evolution as a theory in biology to evolutionism as a materialist philosophy.

This is not really a recent change. Ever since Darwin published *The Origin* and Thomas Huxley defended him against his critics, evolution has been linked with materialism, starting with the writings of Herbert Spencer, a Nineteenth Century philosopher. It was Spencer who introduced the unfortunate (and largely meaningless) phrase "The Survival of the Fittest." But the wide, popular identification of evolution and materialism, at least on this side of the Atlantic, followed the development of the neo-Darwinist synthesis.

There was no single turning point, but particularly public expressions of the change were made during the centenary of the publication of *The Origin of Species* in 1959. At one of the centennial celebrations Julian Huxley said:

"In the evolutionary pattern of thought there is no longer either need or room for the supernatural. The earth was not created, it evolved. So did all the animals and plants that inhabit it, including our human selves, mind and soul as well as brain and body. So did religion . . ." [1]

Here is a scientist making religious statements, not just in a private off-the-cuff way, but at a big international event. On a television programme in the same year he said:

"Darwin's real achievement was to remove the whole idea of God as the Creator of organisms from the sphere of rational discussion."

This is a strange statement. Notice that word "real." The fact that Darwin was a great natural historian doesn't apparently matter. The fact that he proposed a theory to bring together the separate strands of biology isn't important. His *real* achievement was to push God out of the debate! "The sphere of rational discussion" cannot mean science, for science has nothing to say about God. The sentence can only mean that Darwinism has removed the idea of God as Creator from disciplines like philosophy and theology. By what authority can a zoologist make this sort of claim?

Science is a body of knowledge, based on observations and experiment, that can be checked by other scientists. Only things that can be observed or measured can be investigated scientifically, so explanations not based on this empirical evidence are not part of science. Therefore Huxley's statements were not scientific, but philosophical assertions, delivered in a dogmatic manner. None of my tutors at Cambridge would have spoken publicly in this way, and Darwin would certainly not have done so. The references to God and the supernatural are gratuitous and contribute nothing to the body of scientific knowledge. Unfortunately Huxley was not alone in what he said. Other speakers at the centenary made similar comments, and several books were published at about the same time which similarly linked Darwinism to atheism. [2] I cannot understand the motivation of those who made - and still make - such statements. I doubt if they enhance the speaker's prestige in the scientific world, for a scientist is judged by the quality of his work, not his private beliefs.

One consequence of this aggressive atheism is that it invites counter-attack from those who do not share these extreme views. It is not surprising that Christians and other theists saw this as a challenge, and replied with their own facts and arguments. As a result, the debate became intensified.

For many people Huxley's attempts to use evolution to dethrone God was baffling: they had happily accepted God as the prime mover behind the existence of the world and its creatures, and had readily accepted the theory of evolution as a working *scientific* explanation of the diversity of animals and plants. What was the problem? But for others, the enemy was not just evolutionism as a materialist philosophy, but the whole idea of evolution itself, which contradicted the simple account of creation written in Genesis, part of God's Word.

All this time creationism was flourishing in the United States. Creationism and evolutionism had been at war for decades over there, but, in my experience, that did not affect biology teaching here in England. I first came across some American booklets about evolution in the beginning of the 1970s. [3] They were different from the EPM literature: they were written for teenagers, were more chatty, and used the word "creationist" throughout. Hardly any of the older EPM leaflets that I saw were illustrated, but these American booklets had a photograph or diagram on almost every page. The author, Kenneth Taylor, explained that what he was opposing was the assertion of some evolutionists that the "fact of evolution" has freed them from superstitious beliefs about God. A Christian, therefore will want to believe in something other than evolution, if atheism goes along with it.

These booklets were unlike modern creationist literature. The only reference to Genesis was that the author argued *against* the 6 Days of Creation being periods of 24 hours. The Curse (Genesis 3) wasn't mentioned, Noah's flood was dismissed as having nothing to do with the fossils, and the author was content to accept an age for the earth of millions of years.

Later I heard of a new departure called Young-earth Creationism (YEC) through a friend. He was excited about a book called *The Genesis Flood* [4] that had recently been published and which had two key innovations. First was the idea that the earth is just some thousands of years old; a claim supported by a large number of pieces of evidence such as the saltiness of the sea and the depth of cosmic dust. Second was the bold claim that all the sedimentary rocks before the Pleistocene era were deposited during Noah's flood. There was a change of emphasis, from the first three chapters of Genesis to chapters 6 to 8. Of course there were creationists who believed that the world was only a few thousand years old before the publication of *The Genesis Flood*, but YE creationism as a movement dates from around this time.

The approach of the new YEC authors and speakers was different from that of the older Evolution Protest Movement. New topics included the formation of the universe, radiometric dating methods, the Fall of Man and, of course, Noah's flood - things that were scarcely, mentioned by the old anti-evolutionists. The most obvious change was in the tone: these people were on the attack!

Although the evolution sections of English school textbooks were unaffected, this polarisation of views was cropping up in magazines like *New Scientist* and becoming obvious in popular science paperbacks such as those by Stephen Jay Gould and Richard Dawkins.. It wasn't just that people on the two sides of the Great Debate differed in the way they saw things; there was outright hostility and contempt. Consider these phrases, all from the "Letters" pages of *New Scientist*:

"the absurdities of 'Creation Science' . . . "
"the lies of the religious . . . "
"the God Squad are at it again . . . "
"the verbal flatulence that passes for theistic exposition . . . "

(I cannot remember what the actual controversies were that caused these rude outbursts, but I jotted the phrases down out of curiosity.)

The literature published by various creationist groups was nothing like so offensive, but there was a tone of ridicule that was missing from the earlier EPM publications. "Scientists" were often represented as caricature Einsteins with huge foreheads and wild hair. Some of the material was accompanied by cartoons: mad professors, Adam and Eve, piles of bones, joke dinosaurs, and so on. This polarisation has continued and the opposing platforms have become further apart. Creationism began to take the form of "Creation Science" in which the emphasis moved away from the exegesis of the Genesis text to discussion of the "Cambrian Explosion", the Second Law of Thermodynamics and the question of radiometric dating. At the same time several evolutionary biologists became openly metaphysical:

"We take the side of science . . because we have a prior commitment to materialism... That materialism is absolute for we cannot allow a Divine foot in the door."
(Lewontin, 1997) [5]

"Let me summarize my views on what modern evolutionary biology tells us loud and clear . . .There are no gods, no purposes, no goal-directing forces of any kind. There is no life after death . . ."(Provine, 1994) [6]

(This particular statement annoyed me when I first read it! Biology has been my life for over 50 years: how *dare* this man make these foolish remarks in its name? Where is the evidence for these assertions? I was not indignant because these remarks insulted God - it is impossible for a human being to insult God - but because they tarnished biology.)

To some readers these quotations will be old chestnuts. I repeat them not to argue against them, but to emphasise the point that the writers, while claiming the authority of science, have jumped from biology to religion. And their opponents will not let them get away with it, hence the polarisation into hostile camps, the evolutionary atheists and the creationists, each side labelling the other "fundamentalist." This marks a change from the past: the atheism is out in the open. Both ultra-Darwinians and members of the various Creation Science groups claim to wear the mantle of the biological sciences, but the material is transparent and you can see the worldviews beneath.

Consider this sentence from a well-known evolutionist book:

"Darwin made it possible to be an intellectually fulfilled atheist."

(Dawkins, 1986) [7]

This has been quoted so often that it is easy to let it go past without considering how odd it is:

Firstly, it is assumed that being an atheist is a good thing.

Secondly, it misrepresents Darwin, who was distressed when his theory was used to promote atheism. "With respect to the theological view of the question," he wrote to his friend, the biologist Asa Gray, "This is always painful to me. I am bewildered. I had no intention to write atheistically."

Thirdly, evolutionism is such a wide-ranging and elastic philosophy that it has been used to support almost anything, including racism, Marxism and Fascism.

It became clear that in this widening of the Great Debate the two sides were inadvertently promoting and, as it were, "feeding" each other. Fierce atheists ridiculed the creationists and complained about their "attacks on science" but one effect of this was to push people towards creationism. This was partly on the grounds of "If I'm forced to take sides, I know which side I'll take" and partly because people are repelled by incessant rudeness and arrogance. Creationists were not the least bit dismayed by this new publicity and began to hold more meetings and print more books and leaflets. I know the leader of one creationist group quite well, and he is delighted when leading evolutionists attack creationism. When an atheist opponent makes some particularly insulting remark about creationism, the remark is quoted in full, usually without any editorial comment, in the next issue of the creationist magazine. My friend sees such attacks, particularly rude ones, as confirmation that the creationist message is getting across, and he regards them as a challenge.

The dispute was once very different on the two sides of the Atlantic. Stephen Jay Gould wrote that it was as "distinctively American as apple pie and Uncle Sam." [8] For decades there had been battles in the United States about what can be taught in the public schools [9], with legal tussles and famous cases and landmark judgements. "Creation versus evolution" has had a high profile over there. Until recently most British people would not have heard about it. I do not remember ever discussing creationism with any of my fellow biology teachers, nor a parent ever asking me whether we taught evolution in our biology courses. It was much the same outside the classroom. My wife and I have been members of church home-groups for at least forty years, and the creation/evolution issue never arose. Many Christians had never heard of creationism until recently and many have no interest in it. Just before I retired, a Christian physics teacher approached me at the end of a school service and said, "I must have a word with you. I have heard that there are some Christians who don't accept evolution; can you tell me about them?"

The situation is changing fast. In *Rescuing Darwin*, a short book from the public theology think tank "Theos" some recent poll figures are given. [10] In Britain a study found that 17% of those questioned agreed with the Young Earth Creationists that humans were created by God some time within the last 10,000 years, and 37% thought that Intelligent Design was "probably true." These figures are still much lower than those in America, where nearly half the population disbelieves in evolution. Put another way, about two thirds of the sample "believed in evolution", but of those only just over half thought that evolution had removed the need for God. It looks as if many people in this country are either ignoring or reacting against the current ultra-Darwinism.

The Great Debate now has a higher profile. There are references to evolution-and-creation in television programmes and newspaper articles. "Intelligent Design", the most recent anti-evolution development (which is seen by evolutionists as a much bigger threat to their position than creationism) has been the subject of many letters to the papers. Intelligent Design has also now got into the school GCSE syllabus and the text-books (see Chapter 10).

Notes on Chapter 5

1. From Julian Huxley's address in Chicago in 1959. Theistic evolutionists would point out that the second sentence in the Huxley quotation ("The earth was not created, it evolved") is a *non sequitur*.

2. An American zoologist named Simpson, writing about the same time asked the question "Is there a meaning to life? What are we for? What is Man?" and replied that all attempts to answer that question before 1859 were worthless, and that we would be better off if we ignored them completely. What is it about *zoology* that causes its students to come out with these sweeping statements?

3. *Evolution and the High School Student* and *Creation and the High School Student*, both written by Kenneth N Taylor, and published in 1969 by the Tyndale House Publishers.

4. *The Genesis Flood* by Henry Morris and John Whitcomb, 1969.

5. Richard Lewontin, New York Review of Books, 9/1 1997.

6. William Provine, *Origins Research* 16(1/2) 1994.

7. Richard Dawkins, *The Blind Watchmaker*, 1986.

8. Stephen Jay Gould, *Rocks of Ages,* 2001.

9. The words "public school" and "private school" have almost opposite meanings in Britain and America. In England the Public Schools are privately owned and fee-paying. In the USA the Public Schools are the state schools.

10. Nick Spencer and Denis Alexander, *Rescuing Darwin*, 2009 Chapter 3

Chapter 6

The Characteristics of Living Things

During most of my time at Millfield, the teaching of evolution was about changes in organisms: it did not include the "Origin of Life." Our main textbook described evolution in terms of the development of complex organisms from pre-existing simpler organisms over the course of time. Modern textbooks had a page or two on the origin of life; earlier ones did not mention it at all.

It is moot whether it comes under the heading of "evolution" or not. Evolutionist authors today include a section on the origin of life, because they take the NDS to be a unified theory: inorganic molecules giving rise to organic molecules, to living systems, then to cells and then to organisms. In *What Evolution Is*, Ernst Mayr, one of the architects of the NDS, treated the origin of life as part of the evolutionary process.

Before you can talk sensibly about the origin of something you need to know what it is; and there is no satisfactory definition of "life." But you can list life's *characteristics* and somewhere towards the beginning of most older biology courses and school textbooks there was a section called "Characteristics of Living Things". Some of my colleagues used the ugly mnemonic GRRIMEND to help their pupils remember the characteristics. G stood for growth & repair; R, respiration; the second R, reproduction; I, irritability (= sensitivity); M, movement; E, excretion; N, nutrition, and D: Death. You will be glad to know that these characteristics were taught in a more logical order, like this:

Organisms require energy: *Nutrition* is the uptake or synthesis of food (the organic molecules needed for metabolism)

Respiration is the release of energy from the food

Excretion follows from the first two; it is the removal of the waste products of metabolism

This energy is used for: *Growth & repair*; the body enlarges, differentiates and repairs any damage,

Movement of all or parts of the body,

Sensitivity; the body responds to internal and external stimuli

Finally, organisms *Reproduce* themselves

(It always seemed to me illogical to include *Death* as one of the characteristics of Life, but only living things can die, after all!)

At GCSE level, that was about it. We defined the terms, found examples of each, and maybe discussed the difference between a car and a cow, trying to decide whether petrol was the equivalent of food and whether the exhaust counted as excretion. We made the simple, important, point that a car wouldn't work without a man inside it, whereas a cow managed to do everything "on its own." We discussed why the first six characteristics could be copied by man-made things like pieces of machinery, but the seventh, reproduction, could not. But fairly soon we would go on to the next topic.

At A level these famous "characteristics" might be quickly revised at the start of the course, but there was much detailed work to get through, and we didn't linger on them apart from a short discussion about what the words really mean. *Growth* and *reproduction*, for example, are intertwined ideas. 'Growth' can mean different things. If you inflate a balloon it *grows* but there is no more rubber present at the end than there was at the beginning. Living things, however, grow by cell division: big creatures don't have bigger cells, but more cells. The cells have multiplied by *reproducing*, so our original categories are beginning to overlap. (I remember Dr. Picken starting a lecture in Cambridge by saying, "Each one of us is a colony of Protozoa, the cells of which cohere.")

S level was different. We started with the original seven characteristics, and looked for more. The scholars usually came up with four or five extra ones, and, with a bit of prodding on my part, we found we could double the original number, as follows:

Organic molecules

Bodies have their own special chemistry, based on molecules that are not found in the non-living world, such as carbohydrates and proteins.

Nucleic acids

All living organisms have coded genetic instructions in the form of DNA or RNA.

Cells

The cell is the unit of life. We discussed whether viruses counted as living things, and whether whales, midges and bacteria had cells of the same size.

Homeostasis

Living things can control their internal environment in varying degrees. This is obvious in the case of mammals, but is also true of all organisms, even those consisting of only one cell.

Pattern or *symmetry*.

Most living things are of symmetrical shape. We looked at the shells of molluscs and discussed logarithmic spirals. We compared radial symmetry (like that of starfish) with bilateral symmetry (like that of most other animals). If I had known of it then I would have passed on Ken Dodd's wonderful medical summary: "Are you all right? You should have two of everything down the sides and one of everything down the middle." We looked at the symmetry of flowers and leaves and discussed why twig branching is tidy and root branching is messy.

Almost the only way to tell whether you are looking at a fossil, rather than just a lump of rock, is by recognising symmetry or pattern: the coiling of an ammonite, or the repetition of subunits along a piece of vertebral column. I remember when a schoolboy discovered the fossil *Charnia* in rocks in the Charnwood forest near our home in Leicestershire in 1957. It caused quite a stir because it was one of the first large fossils to be found in undoubtedly Pre-Cambrian rock. (We were told on our geology course: "If someone shows you a Pre-Cambrian fossil, either it's not a fossil or else it's not Pre-Cambrian.") *Charnia* was clearly a fossil. But how did we know that it had once been alive? It was because of its shape: it had a symmetrical pattern, with alternate compartments coming off a central 'stalk.'

Living things evolve.

This is true for both meanings of the word *evolve*.

In Victorian times, "evolve" meant "develop"; the word might be used of an embryo. Cells don't just multiply and get larger, they become differentiated. As an embryo develops the original cells give rise to different types of cell: muscle cells, blood cells, nerve cells, and so on.

But organisms also evolve in the modern sense: they change over time from one type to another. This is true, even if you limit it to microevolution. A poodle is quite different from a wolf, or a cooking apple from a crab apple. The majority of biologists also claim that all living things have common descent, i.e. are the product of macroevolution.

Purposiveness.

If you watch a pebble, nothing happens; it isn't *doing* anything. But if you watch a beetle, it is busy; it is up to something. Looking for a mate or moving into the shade are purposeful activities.

The word "purpose" is controversial. Words like *progress* and *success* and *purpose*, when used in biology, may cause hackles to rise. Biologists lean over backwards to avoid being seen as anthropomorphic, and neo-Darwinists are particularly insistent that evolution is not goal-directed, i.e. that it has no ultimate purpose. But common sense tells us that living things *do* show purpose. Professor Medawar, one of the wisest of biologists, once wrote:

"Purposiveness" is one of the distinguishing characteristics of living things. *Of course* birds build nests to house their young . . "[1]

Autonomy.

Living things act on their own. They *do* their living; we don't have to programme them. This quality of "directive self-regulation" is unique to living organisms.

Momentum.

(Not quite the right word, but it will have to do.) There is a vigour, a push, an unstoppability to much of life. If you watch a frog's egg develop (better still, watch a time-lapse film of it) you must be impressed with the seemingly purposeful cell division; the sheer drive within the tiny embryo. (Conrad Waddington once wrote that "a fertilised egg insists, one might say, on changing. The only way to stop it changing is to kill it . . " Aphids and locusts multiply prodigiously, roots crack paving-stones, ants take over a small field, a new volcanic island will become vegetated in a few months, bacteria come to live on bare metal *inside* a suspension-bridge. The language we use underlines this distinction between living and non-living: when plants arrive on a newly formed island we call them "colonisers" or "pioneers": words we don't use for the sand that blows there. It is not surprising that thinkers, like Bernard Shaw, used to write about a "Life Force", although they meant a metaphysical internal power rather than the thing I am trying to describe. There is a fullness and exuberance about living things. C.S.Lewis put his finger on it when he wrote of "the resource, the triumphs, and even the insolence, of things that grow." [2]

The Origin of Life

About life's origin there is much speculation, but an absence of facts. Descriptions of the origin of life are Just So stories (see Chapter 9) with varying degrees of probability. Several mechanisms have been proposed whereby non-living molecules became living systems, but there is no evidence that any of them is right. In a recent television programme the presenter described a "primordial soup" of inorganic molecules and stated that "scientists think that chemical interactions in that soup created the very first amino acids and basic proteins - the building blocks of life. From these ingredients the first primitive cells would eventually emerge." This was presented with computer-generated graphics (and "thrilling" music) giving the impression that it was part of scientific knowledge. An ordinary viewer would never realise that the programme was guesswork from beginning to end.

This mixture of certainty (living things evolved from non-living precursors) and uncertainty (but there is no actual evidence) is shared by "scientific" beliefs that there is life elsewhere in the universe. A splendid example was in a national newspaper the other day: "For we now know . . . that we may well be on the verge of discovering that we are not alone in the universe; that life may not be unique to earth." Compare the cockiness of "We now know" with the emptiness of "may well be on the verge of . . ."

Warm little ponds

Darwin speculated on the possibility of life originating "in some warm little pond with all sorts of ammonia and phosphoric salts, light, heat, electricity etc. present". Experiments on the origin of life usually start off with laboratory models of his "warm little pond". Inorganic molecules are treated in various ways to see what happens. There was a famous experiment in 1953 in which Stanley Miller passed electrical discharges through a mixture of methane, ammonia, hydrogen and water vapour and managed to produce a number of amino acids. Unfortunately these simple organic molecules are so far removed from the complexity of any real living system that they just underline the hopelessness of the biochemical task. As James Le Fanu put it in his recent book,

"The project of itemising the several thousand chemical reactions that take place within the cell does not begin to account for those extraordinary properties that so clearly distinguish the living from the non-living, the animate from the inanimate world." [3]

It seems a hopeless task to try to assemble all the various organic chemicals together into some sort of artificial cell. So why don't the experimenters start with a living system that has recently stopped working (e.g. a dead bacterium) and try to get it started again (i.e. bring it back to life)? You have got the various components almost at the right concentrations and almost in the right place. Would it not be easier to get them going again than to make them from scratch? In one of Hammond Innes' adventure stories [4] some marine salvage men find a deserted steamship, *The Mary Deare*, and manage against all the odds to get the engines going again. Although it was a prodigious task it was a thousand times easier than trying to build a new steamship from scratch. When you think of the number of components in the simplest of known living systems (bacteria have many hundreds of genes, to start with) the problem of assembling something living in the laboratory from inorganic materials is like trying to build a the steamship from pieces of scrap instead of restarting the engines.

This ties in with experiments on bacterial transformation. [5] If live bacteria of strain X are cultured with dead bacteria of strain Y, it is possible for live Y to be produced. The explanation is that DNA from the dead bacteria has become incorporated into the live X cells "transforming" them into Y cells. It seems to be biologically unthinkable that the dead Y bacteria came back to life - but I'm not sure why. Pasteur's experiments in the nineteenth century, which proved that spontaneous generation was impossible were very different. He showed that living things could not appear *de novo* from other organic material. And yet, all discussion about, and experiment on, the origin of life involves the change from non-living to living at some point

Artificial life

Recently a team under Craig Venter has succeeded in making an entirely artificial "chromosome" of synthetic DNA which has been introduced into a living bacterial cell with its own DNA removed. The resulting "synthetic cell" can live and replicate. This remarkable achievement was spoilt by over-enthusiastic reporting. The headline in one daily newspaper, typical of many, was "'We have created life in the lab,' says gene scientist", followed by "Man-made life has been created in a laboratory for the first time."

These statements are untrue and distract from the real importance of the research. Life was *not* created at all: the host bacterium was alive throughout. In another article in the same newspaper a day later, a different columnist set the record straight. He described the successful experiment, then wrote, "As Dr. Venter's less hysterical colleagues have noted, that is not the same as creating life." Unfortunately, he later fell into the same trap as the first reporters by writing, "What has happened to the mystery of life? It has disappeared. There are only molecules and the void . . . .life's essence is just a sequence of chemicals." These statements simply do not follow from this research. It is extraordinary that some people long to be able to state that life is *merely* something-or-other. What is the motive for that?

Self-replication and immortality

There is an idea around that genes are immortal. This is because an organism dies but its genes "survive" in its offspring. This is strange, because genes are not alive to start with. Is common salt immortal, because it cannot die? Is an heirloom immortal because it has survived many generations of the same family?

It is claimed that genes can self-replicate, and that life originated when some molecule or other "learnt" to replicate itself billions of years ago. But there is no molecule known that can copy itself on its own. DNA certainly cannot. Professor Dover writes:

"Genes are not self-replicating entities; they are not eternal; they are not units of selection; they are not units of function; and they are not units of instruction. . . .

"Outside the test tubes of the organic chemist, there is no such thing as a self-replicating molecule in biology and probably there never has been.

"The only free-standing biological entity capable of self-reproduction is the cell . . " [6]

Living bodies, living cells

It is a strange fact that an organism can die but cells from its body remain alive. After a human being dies, many cells of the body stay alive for a long time. Some human cells are kept alive as "cell lines" for research purposes, long after the person they came from has died. Well known are the HeLa cells, still kept alive for research purposes in hundreds of laboratories round the world, which came from Henrietta Lacks who died of cancer in 1951. It is a matter of argument whether such cells can be called "immortal" because they are only kept alive by the use of sophisticated laboratory techniques.

A final oddity

There is still, on a dusty top shelf somewhere in the biology department where I used to teach, a very old guide to the dissection of certain invertebrates. One page has this remarkable instruction:

"*Lightly kill* an earthworm by immersing it in hot water ....."

(my italics)

Notes to Chapter 6

1. from *The Life Sciences* by Peter and Jean Medawar, 1977, pp 11 and 12, italics in the original; cited by Augros and Stanciu in *The New Biology*

2. *Surprised by Joy*, Chapter xii, by C S Lewis

3. from *Why Us?*, 2009, Chapter 9, by James Le Fanu

4. *The Wreck of the Mary Deare*, 1956, by Hammond Innes

5. Griffith's 1928 experiments on bacterial transformation were an important link in the chain that led to the discovery that DNA was the genetic material.

6. From the article "Anti-Dawkins" by G A Dover, Professor of Genetics at Leicester, in the book *Alas Poor Darwin - Arguments Against Evolutionary Psychology* (2001) edited by Hilary and Stephen Rose.

Chapter 7

Evolutionary Developmental Biology (Evo devo)

Evolutionary developmental biology (evo devo) is not new but is relatively unknown. I have looked in several libraries, including biology department libraries, and found no books on evo devo (though a few modern books on evolution mention *Hox* and *pax* genes in passing, without stressing their importance.) But evo devo provides a new way of looking at evolution and makes the NDS and the various forms of creationism appear strangely old-fashioned.

An accessible book on evo devo for the general reader (both in its availability and in the clarity of its writing) is *Endless Forms Most Beautiful: The New Science of Evo Devo and the Making of the Animal Kingdom* by Sean Carroll (see Bibliography).

Evo devo, like most topics in evolution, is difficult to summarise. It is a discipline in biology (particularly animal biology) that compares the developmental processes, from egg to hatching or birth, of different organisms. It studies how embryos grow and develop and how that development is controlled by genetic (and other) factors and thus explores how you get from an organism's *genotype* (its genetic instructions, its DNA) to its *phenotype* (its observed bodily form and workings). At the same time it tries to determine how these developmental processes themselves evolved. This sheds light on the whole of evolution because, as Professor Carroll has pointed out, it is through embryonic changes that changes in form arise.

There are several key concepts in evo devo:

Modularity.

The bodies of animals and plants are arranged into distinct parts, many of which are repeated. Think of the segments of an insect's body or the bones of the spinal column of a vertebrate - or your own fingers. Think of leaves in plants. This modularity is there in the embryo, almost from the start, and it can be compared across different phyla. You can liken this situation to that of a town, which is also modular. Offices, factories, houses, &c., are all housed in "modules" called buildings. These are made of modules such as roofs, walls, doors and windows. The walls themselves are mostly made of modules called bricks. Very different buildings (a town hall and a Fish-and-Chips shop) may both be made of standard bricks. This idea of bricks or building blocks is a useful one: several different authors have compared body parts to Lego models:

"The modular constituents of genes, proteins and many developmental operations, shared across widely divergent taxa, are akin to giant Lego sets. Many supposedly complex adaptations are not the end result of the addition of hundreds of brand-new genes; rather, they are the products of new combinatorial permutations of a limited set of free-floating and widely distributed non-Mendelian molecules.

. . . The Lego construction set can get as big as we wish, starting from a limited set of building blocks . . ." [1]

Kirschner and Gerhart have pointed out that a toy castle and a toy Eiffel Tower, both made of Lego, look completely different from a distance; it's when you come closer that you see how cleverly the same blocks are used for different purposes.

The role of the genes.

It has been discovered that the arrangement and position of these modules in nature are controlled by identical genes in widely different organisms. The impression given by the

NDS is that, during the course of evolution, mutations gave rise to new genes which gave rise to new structures: flies have "fly genes" for making fly bodies and cows have "cow genes" for making cow bodies. Studies of the role of genes during development show that this is too simple, and mostly quite wrong. Development is controlled by "tool kit" genes which are highly conserved across the whole of nature. Genes are switched on or off as the result of the activity of other genes. There is an interacting network of genetic control of great complexity. Diversity is not caused by diverse genes, but by the *same* genes coming under different regulation. As Professor Carroll states:

> "Evolution of form is very much a matter of teaching old genes new tricks." [2]

and,

> "the evolution of form is not so much about what genes you have, but about how you use them." [3]

One of the most remarkable discoveries in evo devo is that these regulatory genes are spread across the whole range of animals so far studied. Famous examples are the *Pax 6* gene (regulating eye development in all animals with eyes, including marine snails, fruit flies and humans) and the various *Hox* genes which are near-universal body-building genes. It is an astonishing fact that the development of completely different eyes (a wasp's eye with its many external lenses and the human eye with its single internal lens) is controlled by the same gene; another tool kit gene controls limb formation in both flies and chickens. Because these genes are found in all the different animal phyla, it is reasoned that they must be of great age: Carroll states that all the genes for building large, complex animal bodies long predated the appearance of those bodies in the Cambrian Explosion.

Developmental plasticity.

Genes interact in highly complex ways; their products affecting other genes. During development a body is continually inventing itself, as well as being shaped and reshaped by genetic and other forces. Phenotypes do not just depend on the genotype. Gene expression is regulated in a complex way and *epigenetic* factors (i.e. non-genic, biochemical or environmental ones) also play a part in that regulation.

Changes of emphasis

When you turn from the NDS to evo devo you move from speculation to fact. No one can ever know for certain what happened in the distant past, but the findings of evo devo about the way bodies are put together, set one's feet on firm ground. Evo devo is much more *explanatory* than the unmodified NDS.

There is a welcome change of focus, away from just DNA to the whole cell and the whole organism, with all their richness and complexity. If evolution was just about genes and their mutations it would be difficult to explain why human beings had so few genes (fewer than twice the number of genes in a fruit fly). Mice and humans have nearly identical numbers and kinds of genes (about 25,000 each). That is an extraordinary fact! It is the *switches* that matter, both in development and evolution, and it seems that they depend on the whole cell, not just its genotype. It is also clear that the *creativity* of evolution is inherent in the living cell or organism and not in natural selection. Selection is dutifully mentioned by evo devo authors, but the real action is elsewhere.

Again, evolution is all about change (evolution *means* change) and yet key words in evo devo are "the same" and "conserved". There is far less change than you would think. Body plans stay the same. There are no new phyla since the Cambrian; fossils stay the same for millions of years. Regulatory genes turn out to be very ancient. Kirschner and Gerhart show that the very diversity of the countless thousands of actual organisms that we see depends on the stability of underlying *conserved* core processes. Neo-Darwinism is often criticised because beneficial mutations are so rare; harmful ones are simply weeded out. But in Kirschner and Gerhart's scheme (facilitated variation) a mutation may cause a wide range of variation which is *not* damaging because it allows a shuffling of flexible and "robust" core processes. Bad changes are shock-absorbed.

These new discoveries have underlined two things, the first of which will not be welcomed by creationists nor the second by strict Darwinists. Firstly, the discovery of ubiquitous regulatory genes is a very strong argument for common descent. Michael Behe has pointed out that if mammals and flies use the same switching genes, it is reasonable to think that they inherited them from the same ancestor or ancestors. [4]

Secondly, evolutionary change need not be gradual. Darwin insisted that it was, and you will remember T H Huxley's response: "You have loaded yourself with an unnecessary difficulty." The discovery of master genetic regulatory programmes for animal body modules allows abrupt, large evolutionary changes to take place. Kirschner and Gerhart refer to evolution taking "large, forceful steps" [5] and a large step is much the same as a jump (saltation). It is not clear why Darwinists are so insistent that evolutionary changes must be gradual. The change from a tadpole to a frog (or from a caterpillar to a moth) is a big jump, but the underlying genes have not changed at all.

The importance of evidence

We are told, forcefully and often, that nothing must be accepted without *evidence*. The reason that evo devo is so compelling is that it is based on laboratory evidence, such as gene sequencing or studies of gene interactions during

embryonic development. Much of the NDS, on the other hand, relies on argument rather than evidence. Take the three neo-Darwinist claims, mentioned in Chapter 4:
1. Evolution is gradual. Most evidence is against this: the classification and the fossil record mostly show discrete groups without intermediates. Evo devo provides evidence that small genetic or other changes can have immediate large-scale results.

2. Mutations are random. There is no evidence that genetic changes in the past were random; it can only be a matter of assumption.

3. Natural selection is the main mechanism. This is a claim with little evidence. Present-day examples are not convincing. The peppered moth, the snail *Cepaea*, the heavy-metal-tolerant grasses, and so on, have not evolved at all; they are all the same species as before. As for the distant past, there is almost no evidence that natural selection was the agent of change. An attempt has been made to correlate the gradual development of the horse's hoof with an increase in wind-blown sand in the rocks surrounding the fossils. The implication was that the habitat was getting drier, and that therefore hooved animals would have a selective edge over plantigrade animals (those that walk on the soles of their feet). But it was not convincing, partly because habitats change often in millions of years, and partly because hooved and plantigrade animals are equally well adapted to the same terrain (think of game animals in Africa). There is no evidence that reptiles gave rise to birds, or that certain reptilian jaw bones became mammalian ear ossicles, *as a result of natural selection*. It's unfair that no evidence is available: you cannot fossilise a process, and present-day natural selection goes too slowly to be studied, and so on . . but the fact remains that the evidence is missing.

That's why evo devo is so important: the evidence *is* there.

Notes to Chapter 7

1. From the article "Anti-Dawkins" by Gabriel Dover in the collection *Alas Poor Darwin* (2000)

2. *Endless Forms Most Beautiful*, page 135

3. Ibid, page 155

4. In *The Edge of Evolution* (2007) by Michael Behe (Ch 9).

5. See *The Plausibility of Life* by Marc Kirschner & John Gerhart

Chapter 8

Innate Transformation

Evolution has come to mean *Darwinian* evolution: evolution-by-natural-selection. When Darwinian evolution is criticised the usual response is "What would you put in its place?" 'Innate Transformation' (IT) is a possible non-Darwinian explanation for evolution. The word *evolution* here means "descent with modification of all organisms from an original primal cell." Like Darwin's original theory, IT is not concerned with the origin of life.

The idea behind IT is that evolution is not something that *happens to* living organisms, but something in which they play an active part. It reflects the original meaning of the word *evolution* (= an "unfolding"), a meaning foreign to the neo-Darwinian concept of random, selection-driven change.

Preliminary argument

For critics of Darwinian evolution the problem is not primarily that of accepting natural selection as the mechanism. (Even young-earth creationists accept natural selection, although they limit it to microevolution). Natural selection clearly happens; it is a description of a fact about the natural world. The problem is where the variation comes from. In neo-Darwinian theory, variation is ultimately caused by *mutation*, and mutation is *random*, but a lot depends on how you define these slippery words. It is impossible to prove that *all* variation is thus caused. IT proposes that heritable variation is innate: internally directed, and caused by genetic, and other, changes that may not be random.

Haeckel gave the name *ontogeny* to the development of an individual from egg to adult, and *phylogeny* to the evolutionary development of the race. He claimed that a human embryo passes through the phases of its evolution, becoming first invertebrate, next fish-like, then reptilian, then mammalian and so on, so that its ontogeny repeats its phylogeny. This idea (Recapitulation, see Chapter 2) was flawed, but a modified form is accepted today. [1] IT suggests that there is indeed a close link between ontogeny and phylogeny but it is the other way round. Instead of evolution shedding light on embryological development, the latter sheds light on evolution. The two modes of development are aspects of the same process and share the same innate mechanisms.

The primal cell was alive. Therefore it must have contained everything needed for life: the organic molecules, the cellular structures and the genes that control them. The genetic material included latent instructions for future organisms in the form, not of actual genes, but of the much smaller genetic modules which make up the genes. In a useful analogy, genes have been pictured as *recipes* for living organisms. A cookery-book contains numerous recipes, each consisting of short instructions such as "add the egg yolks" or "bake for 10 minutes." In living things the full recipes are represented by the genes themselves; the short phrases by the genetic modules. The primal cell contained the "phrases" but not the "recipes" for making future organisms. In the Lego blocks analogy, the primal cell would have a starter Lego kit (still pretty huge, mind) whereas a modern organism would have an advanced kit.

A developing embryo forms as a result of a dynamic interaction between its genetic instructions, themselves fluid and flexible, and its total environment, both internal and external. This is a two-way interaction: the embryo is influenced by its environment, but at the same time helps to create its own environment. The same is true of evolving organisms. Just as a developing embryo goes through stages a, b, c, evolving organisms change from A to B to C. Just as the a, b, c, pattern of development is caused by non-random changes, so might the A, B, C pattern.

In the neo-Darwinian model random mutations are the ultimate source of inherited variation. In the IT model much of the variation is potentially there in the original cell. Mutations increase the amount of variation but in two quite different ways. Some mutations are random in all the various senses of that word. Examples are point mutations and other such mistakes which arise when DNA is copied. Many of these mutations are neutral in their effect, making no difference to the phenotype of the organism. Some are neutral in the sense that, although they have an effect on the organism, it is neither harmful nor beneficial (for example small changes in petal colour in a flower). Many are harmful and cause a loss of information from the genome in the same way that a misprint in a book spoils the sense and results in a flawed book. A few have a positive role in evolution, because they result in a limited advantage, for example fleas losing their wings.

Other mutations are not random, but are triggered by the environment. They cause the cell or organism to respond appropriately to environmental cues. The result might be a heritable change or just a temporary gene switch. These mutations are rearrangements of genetic material to form new genes with new properties. It is these mutations which are responsible for macroevolution.

A mutation is a change in DNA. Organisms also undergo *epigenetic* change; variation that does not involve DNA at all.

A hidden assumption by some who accept the NDS uncritically is that, given time, *any* mutation might crop up, *any* organic molecule might be synthesized or *any* body plan might come into being, with natural selection acting as a filter for all these countless possibilities. This is not so. There is a limited number of viable forms available for any organism (an estimate of fewer than 1000 has been quoted for animals), partly based on what is mathematically or physically possible. [2] (This is why creators of science fiction struggle to come up with convincing monsters.) There is a limited number of genetic modules, and a limited number of available organic molecules which can be used in a working body. So the same molecules pop up all over the place, appearing in very different organisms with the same, or different functions. Haemoglobin is found in earthworms and some plants as well as vertebrates; trypsin is involved in both digestion and blood clotting; the juvenile hormone of insects is also found in yeasts, soya beans and some human organs. A good example of a multipurpose molecule is adenine, present in the hormone cyclic AMP (found in both humans and slime moulds), the energy carrier ATP and the building blocks of DNA. You can see why the Lego analogy is a good one for living structures: new models are not necessarily made with new bricks, but with rearrangements of old bricks.

There are mathematical, mechanical and chemical limits to the range of forms and physiologies available to living things. Because of these limits, the idea that phylogenetic pathways are innate becomes less improbable.

Direct evidence for Innate Transformation

Genetics

Advances in biology have demonstrated that genes are modular, that the modules can be shuffled and transposed between chromosomes and that the development of an organism is a set of unique interactions between its genetic instructions and the environment. [3]

Body plans

There is evidence that living things have a limited number of basic designs which are governed by the same regulatory genes. Beetles are not the product of "beetle genes", and horses of "horse genes", but both result from the *same* genes doing different jobs. (Examples already noted are the Hox genes which control the basic body plans of all animals and the pax 6 gene which organises the construction of the eye in all sighted animals. These genes are known to be ancient. [4]

Non-random mutations

There is experimental evidence that evolution can be the result of mutations which are not random, but triggered by environmental cues. [5]

Indirect evidence for Innate Transformation

Orthogenesis

An embryo shows trends (is "directional") during its development. Organs form in an orderly way; blood vessels or limbs elongate progressively. Evolving organisms also show trends through time. This progressive evolutionary development has been called *orthogenesis* ("evolution-in-a-straight-line") and it is a feature of the fossil record. The best examples are those fossils which form a *series*: i.e. which show directional changes in progressively younger rock strata; for example the graptolites in Cambrian, Ordovician and Silurian strata which show a trend from being many-branched to being single-branched. Series of starfish and sea urchin fossils are found in the chalk of the Cretaceous. There are trends within small groups (orders or families) such as the increase in complexity of the suture-lines in the ammonites, or the progressive changes seen in the horses. There are bigger trends across phyla or even kingdoms: for plants to become vascular, say, or for amphibia to become reptilian. These trends are hard to explain on the NDS but would be expected under IT. Take the trend for reptiles to become mammalian, for instance. Towards the end of the age of the dinosaurs, a number of different groups of "mammal-like" reptiles appeared in the fossil record, some not obviously directly connected with any of the others, nor with the true mammals that followed them. Characteristics such as the details of the jaw hinge, the dentition, the position of the limbs with respect to the spine and so on seemed to become more and more mammalian in group after group. It was as if "mammalness" in general was being evolved, rather than descent through one line of ancestors. This is evolution, but it is not Darwinian.

The Cambrian explosion

At the beginning of the Cambrian period, about 550 million years ago, the rocks become suddenly full of fossils. This is after thousands of millions of years of Precambrian rocks with almost no fossils at all. Stephen Jay Gould described the Cambrian as the crucial time when multicellular life first appeared in a "geological whoosh." [6] and as the "key event" in its history. [7] Almost all the major phyla of animal life appear, apparently without predecessors It is a *fossil* phenomenon. If it wasn't for these Cambrian fossils there would be no reason to think that all the different phyla of invertebrates had appeared at much the same time in life's history. Nearly all fossils are the remains of the hard parts of an animal, such as shells and skeletons, so this was an "explosion" of hard-part-formation, as well as of new body-plans. Now, just as the skeleton begins to form in an embryo in many different structures at the appropriate time, so hard parts formed in many different phyla during the Cambrian. It is possible that the appropriate genes were assembled or switched on when minerals in the sea reached a certain concentration.

Saltations

A saltation is a sudden event (literally, a jump). During embryonic development there are many such changes, for example the development of limb buds, the opening of an orifice or the start of the heartbeat. According to the NDS there should be no jumps, or large gaps, in evolution. If there is an apparent gap, biologists look for an intermediate form - or "missing link" - to fill the gap. Whether or not there is such a link in any particular gap is a matter for discovery (fossil) or research (genetic analysis). But a gap does not necessarily have to be filled. If the analogy between embryology and evolution is correct, to insist on a missing link between two phyla might be like wanting a transitional structure in the body between the stomach and the liver. Again, to say "men came from monkeys" might be like saying "the heart came from the lungs." Actually, in anyone's book, the phrase "men came from monkeys" has to be wrong. "Men and monkeys had a common ancestor" is nearer the mark, as is "the heart and the lungs are derived from a common embryonic tissue."

Convergence

Convergent evolution results in unrelated organisms having shared characteristics. Penguins and fish both have streamlined bodies. Opossums (marsupials) and spider monkeys (placental mammals) both have prehensile tails. Cacti and Euphorbias, although in different plant families, both have thick waxy cuticles and swollen, water-storing, stems. The animals and plants in these pairs are not closely related. Convergence is hard to explain by the NDS. Firstly, there is the unlikelihood of the environment being fine-tuned enough to produce the extreme similarities between unrelated but convergent species that are actually found. Secondly, how can *random* mutations in quite separate lines of descent give almost identical results? [8]

Thirdly, the findings of evo devo have put something that appeared clear-cut back into the melting-pot. Can structures be said to have evolved *independently* when they are governed by the same regulatory genes? If IT is true, convergence would be normal and unremarkable: similar environments would trigger the same responses from identical genetic modules in organisms from widely different parts of the classification.

The speed of evolution

The evolution of some organisms has been too rapid to be accounted for by the NDS. Here is an example from the mammals: A chronospecies (fossils that have changed so little over time that they are judged to be still the same species) is estimated to last, on average, for about a million years. Now bats and whales are supposed to have evolved from a common ancestor during a period of about ten million years. But that means there are only ten to fifteen chronospecies between a primitive mammal and a bat or a whale, which is impossible if the changes are random. [9] However, if IT is correct, evolution can be hundreds of times faster because the random element is reduced.

Mosaic evolution

Different structures in an evolving line of organisms usually change at different rates in a piecemeal fashion. This is known as "mosaic evolution" and is characteristic of transitional organisms, i.e. organisms thought to be on the ancestral line between an earlier and a later form. *Archaeopteryx* is a good fossil example, having many reptilian features (teeth, long tail, clawed forelimbs) and avian features (beak, feathers, large brain). The platypus is a good living example because it is a furry mammal that lays large, shelled eggs like those of a reptile. These are very strange creatures. It's as if *Archaeopteryx* was trying to become a bird and the platypus was trying to become a mammal and they had only got half-way. They shed no light on *how* wings evolved, or *how* mammals acquired their unique method of reproduction. Mosaic evolution is explained genetically by the independence of regulatory genes. This is good evidence for IT: embryos also exhibit mosaic development with different structures developing at different rates.

Epigenetic landscapes.

Conrad Waddington, writing in the 1950s, illustrated the development of an embryo with the image of marbles rolling down a hill. There are grooves or valleys running down the hillside which branch and multiply as you go further down the slope. The various parts of an embryo develop into different tissues and organs by moving down a similar "epigenetic landscape", following the easiest routes down the floor of the valleys. The hills and valleys correspond to the various genetic, chemical and mechanical constraints on development. Once the "marble" (an organ, say) has passed a junction it cannot return; its fate is fixed. However just as a small nudge at a junction could send a marble into a different groove on the sloping hill, a small biochemical 'nudge' can change the fate of a tissue or organ. An early nudge, although a small change, can have a large effect on the end result. The nudges during development of the embryo may be innate (caused by the products of regulatory genes) or external, provided by the embryo's local environment.

This provides a good analogy for IT. The course of evolution can similarly be pictured as a flow down an epigenetic landscape, with comparable innate or environmental 'nudges'. (This also explains why evolution is irreversible.)

No new phyla have appeared in evolution since the Cambrian period (roughly 500 million years ago), no new classes since the Silurian (roughly 400 million years ago) and no new orders since the end of the Mesozoic era (about 60 million years ago). At the same time, the number of species has multiplied. This is the pattern you would expect if there was an evolutionary landscape analogous with the embryological epigenetic landscape. It is possible to draw a diagram with a pattern of branching lines to indicate *cells / tissues / organs / systems* which resembles the equivalent diagram drawn to indicate *species / genera . . . orders / phyla* (except that the latter has more divisions). Notice that when they first appear, the higher categories - organs, phyla, and so on, are just cells or groups of cells; potential organs and phyla rather than actual ones.)

The branching-places in the epigenetic or evolutionary landscapes, the junctions where the 'nudges' take place, are tiny and local, but the results are far-reaching and irrevocable. A fork in the road is a small thing, but if you took one wrong turning when driving out of London you might end up in Cornwall instead of Kent.

Family 'trees' have been constructed to show the course of evolution. Many of them depict an actual tree shape, with a thick single trunk at the bottom, then branches, then small twigs at the top. This is a misleading image, especially with regards to the big single trunk. A better illustration would be of a sprawling seaweed with many thin stalks (the phyla) coming off near each other low down, followed by multiple branching at various levels and culminating with millions of separate ends. There is only one diagram in the whole of *The Origin of Species*, a simple line drawing of a small part of a family tree, which must have been a favourite of Darwin's because he kept referring to it. [10] If you turn any of these tree diagrams upside down you will see that they are the same shape as the epigenetic landscape. Darwin called evolution "descent with modification" but his diagram depicted *ascent*, i.e. an upwardly growing tree. These words are used interchangeably. (In 1871 Darwin wrote a book called *The Descent of Man* in which he attempted to describe human origins. A century later J.Bronowski made a series of TV programmes called *The Ascent of Man*, on the same subject.) In terms of the mechanism of evolution *descent* is the better idea: an epigenetic landscape is more apt than an evolutionary tree. Writers on evo devo use the word *descent* over and over again.

Recent work on the sequencing of DNA shows that the idea of a simple tree-of-life must be abandoned. If genes could only move in one direction, from the trunk to the tips of the branches, a tree image makes sense, but one-celled organisms can exchange DNA "sideways" (horizontal gene transfer). Evolution in both plants and animals may also have involved hybridisation. [11] No tree has lots of horizontal connections between the branches and twigs. But turn the picture upside down and the image resembles a river delta, where lateral connections between the main channels are frequent.

There is another difference between a tree diagram and that of an epigenetic landscape. Tree diagrams are of just the tree, growing up into a conceptual vacuum. In a diagram of an epigenetic landscape, there are *two* separate things shown: the rolling marble and the series of channels. The marble represents the changing living structure and the channels represent the constraints, or the environment around the living structure. Both are important. [12] Most emphasis in the NDS has been on the genes (part of the "marble") but the channels are just as important in development. This too relates back to Darwin, who never abandoned the belief that the environment played a causal role in evolution.

Exploratory processes

If you look at a colony of amoebae, or a group of cells in a tissue culture, they can be seen to be vigorously moving about or putting out bits of themselves (pseudopodia) in order to explore their surroundings. This exploratory activity is so widespread that it could be included in the list of "Characteristics of Living Things" (Chapter 6). The roots of vascular plants explore the soil by probing and branching, "searching" for water or nutrients. Fungal hyphae spread in the same way, pushing through the food supply. In an animal embryo the cells move about, assemble and arrange themselves into the various tissues. In particular, the precursors of the blood vessels and nerves advance towards their target areas, forming and reforming in an exploratory way. (Incidentally, this is why the insides of animals look "messy": the messiness, like the untidiness of a root system, is a consequence of the exploratory processes of development. The embryo makes itself, trying this and trying that, developing its own unique form. If I were to take part in the "Design" debate, I would argue that these exquisite responses to local needs and conditions were actually better evidence for Design than a fixed, formal arrangement. The "Designer" is further back in the process.)

The idea behind IT is that evolution is analogous to embryonic development. Is it possible that evolution itself is exploratory? If you look at the family trees of the classification with an open mind it is possible to see the pattern of theme and variation as an exploratory pattern. This is reflected in the way people speak: the beetles have "come up with" different types of camouflage; tarsiers have "arrived at" a nocturnal way of life. A good example of such an exploratory process is Adaptive Radiation (remember the marsupials in Australia) in which a new ecosystem is explored by evolving organisms.

Arguments against Innate Transformation

IT is based on an analogy between evolutionary development and embryology, both starting from a single cell. This is a false analogy because the cells have such different origins. A fertilized egg is the product of two sex cells from highly evolved parents, whereas the "original" or primal cell is not.

Agreed; at this point the analogy breaks down. But it is only an analogy and other parts of IT have explanatory power. It is worth pointing out that *nothing* is known about the primal cell except that it was alive.

(Note: this exact analogy was also used in the exchange between J.B.S. Haldane and the skeptical woman who didn't think it was possible to go from a single cell to a complicated body even in the billions of years allowed for evolution in theory: "But madam, you did it yourself. And it only took you nine months.")

It is preposterous to endow a primal cell with the genetic capabilities of highly evolved organisms.

Not really. It depends what you mean by capabilities. The primal cell was alive, and therefore must have possessed all the structures and biochemistry needed for life, including genetic material. A similar criticism about the genetic content of the primal cell has appeared before. In his book *Darwin's Black Box*, Michael Behe speculated that the first cell already contained the irreducibly complex biochemical systems that would be found in later organisms. (He referred only to "systems" but in the context these would include the genetic instructions for constructing them.) In *Finding Darwin's God*, Kenneth Miller attacks Behe's suggestion, pointing out that the idea of a bacterial cell being full of genes that would not be turned on for billions of years is ridiculous. Miller's criticism makes sense if *genes* are insisted on, but the existence of gene modules is another matter entirely. The facts of biochemistry are overtaking this argument.

IT theory, like orthogenesis, is just a form of vitalism, and is thus as suspect as cosmic teleology (i.e. purposeful, 'goal-directed' evolution) or Intelligent Design.

It depends what you mean by "vitalism". A. G Cairns-Smith described vitalism as a ". . deeply mysterious unifying power, a principle of life, an essential magic about living things that divides them from everything else." [13] But why use those loaded words "mysterious" and "magic"? It is a fact that living things *are* divided from everything else.

IT brings in nothing from outside biology. If vitalism means that cells are alive, then, yes, they are vital. Living things are alive and no one knows why, nor has life been adequately defined, let alone understood. Living things are autonomous: they divide and reproduce and embryos develop.

And living things evolve. The process is innate, but it is moot whether you describe it as a *drive* or not. When you think about it, natural selection only operates because organisms are alive. Is that, too, a form of vitalism?

Orthogenesis is disallowed by mainstream evolutionists. Ernst Mayr called it a "refuted hypothesis" and claimed that it was wrong because it was purposeful in some metaphysical way. [14] But orthogenesis has not been refuted: it is supported, if anything, by the fossil evidence. It is rejected because it is philosophically unacceptable to defenders of the NDS. In his recent book on fossils Brian Switek described orthogenesis as the hypothesis that evolution was "striving towards particular goals." [15] That is the NDS view of orthogenesis, but "striving" and "goals" are loaded words and put up a straw-man target to attack.

Here's a curious point: neo-Darwinists regard fossil *series*, such as the famous horse series and the series of human "missing links", as powerful evidence for evolution. Such series establish both that evolution happens and that it is gradual. Now a series implies a *trend*, almost by definition. But, perhaps on a different page of the book, trends are denied because they imply orthogenesis. You cannot have it both ways.

It is useful when considering accusations of vitalism or teleology (or anything else hinting at metaphysical naughtiness) to go back to the living cell or the developing embryo and ask some questions. Is respiration purposeful?

Is cell-division innate? Is embryology goal-directed? Does a chick have a drive to escape from the egg? The same questions can be asked and answered with respect to evolution. If IT is teleological, so is embryology; if embryology is not, neither is IT.

The idea of IT simply pushes the main problems of evolution back to the primal cell.

That's where they have always been. The Origin of Life is a much more difficult topic than Descent with Modification. Whether by studying actual organisms or arguing theoretically, it is much easier to derive a mammal from an *Amoeba* than it is to derive a living cell from a handful of molecules. Firstly, there is a frontier to be crossed: cells are alive whereas molecules are not. Secondly, there is no empirical evidence for any change from non-living to living.

The timing is counter-intuitive; why the millions of years?

Why does anything take a particular time? Some processes in biology, like the electrochemical events in nerve cells, take milliseconds. Others, like the growth of a bristlecone pine tree, may take thousands of years. If IT is right, it is likely that at least some of evolutionary timing is the result of large-scale, extremely slow, planetary changes. For example, certain genetic switches could not be thrown until the oxygen in the atmosphere reached a minimum concentration.

The IT model is based on the analogy between the development of an individual and the history of the race. But the former has an end-product, the adult organism.

The comparison between ontogeny and phylogeny is over a century old and has always had a measure of explanatory value, end-product or not. Secondly, the term "end-product" may not be appropriate for living things, which are always at some particular stage in a series of continual life-cycles.

Thirdly, here is a provisional possibility: imagine an animal going slowly through many larval stages, A...B...C... An investigator finding B for the first time, not knowing about A or C, would accept it as a perfectly good organism, standing alone, as it were. Later on, when he knew the whole life history he could see A, B, C as stages in a process. In evolution, A, B, C and the rest are not individual organisms but aggregations of the total living world at a particular time, and the geological column is a series of cross-sections through these developing biospheres.

Here is a lateral thought. During evolution, countless lines of organisms have come to an end. There are no more trilobites, ammonites or dinosaurs in the world. Traditionally, there are two reasons given for this. The first is that the organisms in question lost the struggle for survival when better adapted competitors appeared on the scene. The second reason could be loosely called bad luck. A forest fire might destroy the last survivors of a particular race, or a whole population might die out in one of the cataclysmic extinctions that have punctuated the fossil record.

Might there not be a third reason? Just as an individual gets older and weaker and eventually dies, might not the same be true of an evolutionary line? Could a whole race "get tired", and, as it were, lose the will to live? I can't help thinking of the ammonites. The last few genera which survived into the Upper Cretaceous almost entered a second childhood with their suture lines reverting to a simpler pattern and their shells beginning to uncoil.

Embryological and evolutionary "development" are completely different, so there is no sense in comparing them.

This is a telling criticism. It is the Medawar argument, clearly stated in his essay on Herbert Spencer. Spencer, back in Darwin's time, had argued that embryological development and evolution in the modern, Darwinian, sense were both part of a universal principle of evolution in which undifferentiated things become differentiated, simple things become complex, and so on. But Medawar stated that a frog's egg is more highly evolved than, say, a grown-up earthworm. He went on to argue that it was impossible usefully to compare embryonic development with evolution, because they are "altogether different phenomena." [16]

But are they *altogether* different? They are linked together by the chemical or cellular instructions underlying these processes. Embryological development involves *carrying out* the instructions and genetical evolution involves *changing* the instructions, true, but the picture is altered if we take our eyes off DNA for a moment and look at the whole cell. DNA doesn't actually *do* anything, any more than any set of instructions (a blueprint, or a recipe) does anything. The living cell is the agent. It is the cell which reads the instructions. We must look at the reader, as well as that which is read. Many of the processes of development in the embryo are independent of the DNA 'instructions'. The development of capillaries, the branching of larger blood vessels, the actual innervation of tissues and organs, the formation of trabeculae in bone: these things are exploratory and depend on non-genetic factors. In the same way the fine details of structure of actual organisms and the way they pass down from generation to generation may be nothing to do with genes and their mutation. Life (whatever it is) is transmitted through the cytoplasm and not via the DNA, and both forms of development, embryonic and evolutionary, are manifestations of life.

Notes to Chapter 8

1. Embryos do not repeat the adult form of their ancestors, but they retain certain characters of the young stages of those ancestors. For example, all land vertebrates have gill arches early in their development. In reptiles, birds and mammals, these never become actual gills, but they give rise to other structures in the neck region. Other ancestral structures are retained because they act as 'organizers' which regulate later steps in development.

2. Cells and bodies will tend to have structures which are optimally efficient. See *On Growth and* Form by D'Arcy Thompson.

3. See the article 'Anti-Dawkins' by Gabriel Dover in *Alas Poor Darwin*

4. Gabriel Dover (ibid); *What Evolution is* by Ernst Mayr (transposable elements, *hox* genes, *pax* genes etc; *The Edge of Evolution* by Behe (Chapter 9).
The best popular source for all this information is Sean Carroll's book *Endless Forms Most Beautiful - The New Science of Evo Devo and the Making of the Animal Kingdom* (2006).

5. Spetner, *Not by Chance* (Chapter 7)

6. Stephen Jay Gould, 1993, *Eight Little Piggies* Chapter 15. He does explain that this geological whoosh is a few million years long!

7. Gould, *Dinosaur in a Haystack*, 1996 (Chapter 9)

8. Spetner, *Not by Chance!* (Chapter 4)

9. Data from *The New Evolutionary Timetable* by Steven Stanley

10. *The Origin of Species*, Chapter IV.

11. See *New Scientist* of 24th January, 2009: "Uprooting Darwin's Tree" by Graham Lawton,

12. What if the "landscape" is flexible? There are huge possibilities here.

13. *Seven Clues to the Origin of Life* (Chapter 1) A G Cairns-Smith (1985)

14. See *What Evolution Is* by Ernst Mayr. He dismisses orthogenesis in several places in this book.

15. *Written in Stone* (Chapter headed "From fins to fingers") Brian Switek

16. Sir Peter Medawar *The Art of the Soluble*, 1967 ("Herbert Spencer and the Law of General Evolution")

Chapter 9

Stories and Just So stories

You will not have heard of Innate Transformation: it is something that I put together while preparing the contents of this book. I was sitting on a boulder full of fossil ammonites on Monmouth Beach at Lyme Regis, thinking about geology in general, and fossils in particular, and wondering what their history really was. So little is known about the remote past. There is some fragmentary evidence, but the facts are scarce. So scientists and historians put together *stories* to account for the few facts. Some of them are probable, and some are fanciful. The probable stories are those that make good sense of what evidence there is. If new evidence confirms them, they become accepted as part of scientific or historical knowledge. Continental drift is a good example of a story that is now accepted, although it was ridiculed to start with. The less probable stories are often called "Just So Stories" after Rudyard Kipling's famous tales for children.

On Monmouth Beach I was thinking about the evidence for evolution from embryology, and wondering how much truth there was in the idea of Recapitulation. What if Haeckel was holding the truth upside down? He claimed that the embryological development of an individual organism repeats the evolutionary history of the race. But what if it was the other way round? So I devised my own Just So story and called it (provisionally) "Innate Transformation."

It is a mistake to think that a story, even a Just So story, is somehow trivial or childish. In *The Art of the Soluble* Professor Medawar wrote:

"What exactly are the terms of a scientist's contract with the truth? . . . a scientist, so far from being a man who never knowingly departs from the truth, is always *telling stories* in a sense not so very far removed from that of the nursery euphemism - stories which might be about real life but which have to be tested very scrupulously to find out if indeed they are so."[1]

and in a footnote he explained that a story was more than a hypothesis: it is a theory plus what follows from it and goes with it; a complete 'package.'

Because evolution refers to the past it is a rich source of biological stories. You could argue that it consists entirely of stories (Darwin used the phrase "we may suppose" countless times in *The Origin of Species*). How did life begin? How did flight evolve? Where did flowers come from? Well, no one knows, but there are plenty of stories. They are unlike the facts of hard science because, although some are well-established, they cannot be *proved* to be correct. Not just fairy-stories but *every* story about the past should begin with the words, "Once upon a time . . . "

Schoolmasters know that telling stories is one of the most effective ways of teaching. Biology is rich in marvellous tales. How a seedling turns into a tree; how sand dunes can turn into forest; Mendel's work on peas in the monastery garden; the gradual discovery that DNA is the genetic material: these are all exciting stories, worth telling for their own sake, not just because they are part of a biology course. We had some splendid videos to supplement our teaching. Long after our pupils have forgotten the details of the syllabus they will remember David Bellamy "doing the Indian rope-trick with a tree"[2] and the exact moment when Watson and Crick made their breakthrough in the discovery of the structure of the DNA molecule. Some parts of evolution are dull, or difficult for pupils to understand, but a story can bring them to life. "Have you heard about Bumpus and his sparrows?" is much more interesting than a reference to stabilizing selection. The same goes for Johannsen and his beans, Kettlewell and his peppered moths and Darwin and his finches.

A "Just So story" is a special type of far-fetched story. The original *Just So Stories* were written by Rudyard

Kipling for children. [3] Some of the stories form a sort of nonsense evolution: "How the Camel got his Hump", "How the Leopard got his Spots", "The Elephant's Child" (how the elephant got his trunk). (A video we used to show at Millfield contrasted Lamarckian and Darwinian evolution ("how the giraffe got his neck") by parodying Kipling in cartoon form, even using his language: "Once upon a time, O Best Beloved, there were some giraffes feeding on the forest trees . . . ")

Some people use "Just So story" as a term of ridicule; others use the phrase for something reasonable but unproveable. The claim that *Archaeopteryx* is transitional between a particular group of reptiles and a particular group of birds has been described as a Just So story, on the grounds that you can never demonstrate that a given fossil was the descendant or ancestor of any other given fossil. A Just So story may be the best that anyone can do. "How did reptiles evolve into birds?" is a question that has attracted widely different Just So stories (they ran along the ground and took off . . . ; they glided down from trees . . .) each with its followers.

Some of these stories are pretty daft, like the theory that flamingos evolved their pink colour so that they could not be seen against the setting sun. But some outrageous stories (like continental drift) turned out to be correct. Just So stories are often preposterous. Could anything be more absurd than the idea that human beings evolved in a few feet of sea-water? That continents can slide about the world? That a non-living molecule could be called selfish? That a bear could evolve into a whale? All these things have been suggested, mostly by scientists. Most of them are wrong. But their rightness or wrongness does not depend on their likelihood or absurdity, but on whether there is any evidence to support them. In the case of continental drift, Wegener's ideas were originally rejected because there seemed to be no *mechanism* by which continents could move about. Plate tectonics has now provided that mechanism.

This mirrors what happened with the great-grandfather of Just So stories - the theory of evolution - itself. People played with the idea of evolution for thousands of years, but it wasn't until Darwin, and others around the same time, hit on a possible *mechanism* - natural selection - that evolution began to be generally accepted.

Some areas of modern evolution theory are particularly prone to a multiplication of Just So stories, because they are nearly evidence-free. First is how life began. I recently looked at a book in the library entitled *The Origin of Life* and it was Just So story from cover to cover. It was authoritative but imaginary; ostensibly scientific but with no facts that had any bearing on the title. Another area is evolutionary psychology, which attempts to derive how we think and behave today from our prehuman forebears via early hominids. We are told that a team of scientists has done extensive research to show that men are better at reading maps, or that girls prefer the colour pink *because* . . . and there follows a Just So story about cavemen. It is impossible to know whether these stories - some ingenious, some silly - are right or wrong. There is no way of finding out. [4]

There is nothing wrong with Just So stories, as long as they are not given the prestige of hard science. However Professor Thompson took a much sterner view in his *Introduction to the Origin of Species* in which, having admitted that Darwin's famous work had very greatly stimulated biological research, he went on to state:

"A long-enduring and regrettable effect of the success of the *Origin* was the addiction of biologists to unverifiable speculation. 'Explanations' of the origin of structures, instincts, and mental aptitudes of all kinds, in terms of Darwinian principles, marked with the Darwinian plausibility but hopelessly unverifiable, poured out from every research centre." [5]

The key words are *plausible* and *unverifiable*. These are useful words to keep at the back of your mind when you are reading about Origins.

I recently read three modern books on evolution; by Wilson (*The Diversity of Life*), Carroll (on Evo Devo) and Kirschner and Gerhart (on Facilitated Variation). The books are ostensibly about evolution but the interest was in the actual biology: the description of tropical ecosystems (Wilson) or the new discoveries in developmental biology (the other authors). I read the books to learn more about evolution, but it was the current biology that was fascinating. Why were the "evolutionary" paragraphs less interesting than the rest? Surely because the biological facts are real whereas the evolutionary ideas are speculation.

Meanwhile, what about Innate Transformation? Whether it is plausible or not is a matter of opinion. But I suggest that it is partially verifiable. It is a safe prediction that more ancient regulatory genes will be discovered. (These will link all the kingdoms, not just animals.) It is reasonable to predict the discovery of more non-random mutations. I suggest that the concept of a gene will change; that the gene will lose its unique importance. Whole cells, or parts of the cytoplasm, or genetic modules or even pieces of junk DNA will turn out to have a part to play in inheritance and evolution.

Teachers and text-books traditionally use the famous, flawed, "Giraffe Story" to explain the difference between Lamarck's idea of the inheritance of acquired characters and Darwin's idea of evolution by natural selection. Let us add IT and see what happens:

The Lamarckian explanation

The habit of continually reaching up for food caused the giraffe's neck to become stretched. These longer necks were inherited and over thousands of generations the modern long-necked giraffe evolved.

(Arguments against: (a) It is impossible to stretch your neck (muscles can only contract, not push); (b) acquired characters are not, in fact, inherited.)

The Darwinian explanation

Populations of "early" giraffes included a wide range of neck lengths some of which were genetically determined. Those with longer necks could feed on vegetation that was higher up and thus exploit a food source denied to their shorter-necked fellows. In the struggle for survival the long-necked forms were thus more likely to reach maturity and pass on any (long-neck) genes. After thousands of generations, the result of this natural selection is the modern giraffe.

(Arguments against: (a) The story about the height of the vegetation is unconvincing (what about the dozens of species of small grazers and the juvenile giraffes, who manage perfectly well?); (b) unless there is some form of orthogenesis, or non-random mutation, involved there does not seem to have been enough time to allow for these changes to have taken place.)

The IT explanation

In any appropriate ecosystem there will be a radiation of animal forms, whose body plans result from the interaction of a flexible genotype, innate genetic and epigenetic regulation and external environmental cues. In the hooved mammals, there will be a whole range of possible sizes, appropriate to the particular ecosystem. One animal happens to be the tallest - the one we call the giraffe. It has a remarkably long neck to match its long legs; otherwise it would find drinking, and eating low vegetation, very difficult.

Notes on Chapter 9

1. From the essay "Two Conceptions of Science" in *The Art of the Soluble*, 1967, by Peter Medawar.

2. David Bellamy repeated a famous experiment, first performed by Strasburger. Bellamy and his team supported a large forest tree with scaffolding and ropes, then cut out a section of the trunk, a few feet up from the ground. They put a bath under the cut end and filled it with water, which the tree drew up. They added picric acid to the water to kill the cells of the tree. The tree continued to draw up many gallons of water, even though it was dead.

3. Rudyard Kipling, 1902 *Just So Stories for Little Children*

4. See *Alas Poor Darwin* edited by Hilary Rose and Steven Rose for a critique of some examples.

5. W R Thompson, 1956 *Introduction to the Origin of Species*

Part 2 The Nature of *Genesis*

Introduction

Part 1 concerned the scientific evidence for evolution and natural selection. Any anti-Darwinist books referred to were by authors (mostly scientists) who accepted evolution.

Part 2 ranges more widely because The Great Debate is about everything, not just science. To consider creationist claims, for instance, it is necessary to compare scientific findings with the Creation Account in *Genesis*. This needs a separate section: "religious" ideas are (nowadays) inappropriate for a scientific discourse. It isn't a question of Truth, but of viewpoint and even vocabulary. There is a dislocation half-way through both of Michael Behe's two books on Intelligent Design (*Darwin's Black Box* and *The Edge of Evolution*) when he turns from describing intracellular biochemistry to discussion of a Designer. Note: I am *not* saying that Behe is wrong; there may be no way round this. But the dislocation remains.

This awkwardness is a modern, certainly post-Darwinian, phenomenon. Until mid-Victorian times scientific and religious ideas were juxtaposed happily in scientific literature and books on natural history. I own an early Victorian book on insects which is illustrated with engravings. Here are some of the titles to the illustrations:

Silk Manufacture in China
Muscles of the Cockchafer
Various Chrysalises
The Gnat's Boat of Eggs
St Paul Preaching at Athens
Perfect Insects

St. Paul's sudden appearance amongst the cockchafers and chrysalises is startling to a modern reader who would think it inappropriate in a book on entomology. It isn't that there is no connection between biology and religious faith. *Of course* there is: all truth is interconnected. But nowadays we are unused to the juxtaposition of scientific and religious matters, partly because science writing today uses technical or analytical language, and partly because secularists try to drive a wedge between them.

Evolution theory is part of science, but the 'isms' (evolutionism, Darwinism, creationism, materialism) which surround it are not. They are worldviews. Part 2 is about these, and how they relate both to science and to Christian belief.

In the introduction to Part 1 I described the Nuffield Filter: a useful test to apply to the confident assertions so loved by those on both sides of the Great Debate:

Is this certainly true?
Is this probably true?
Is this possibly true?
Is this false?

To introduce Part 2 here are some more sentences, all taken from actual published work, for the reader to try the filter on:

a) "Life occurs automatically whenever the conditions are right."
b) "Evolution and theism are completely incompatible."
c) "Somewhere in our DNA reside the differences between ourselves and apes."
d) "The moral catastrophe that has disfigured modern Western society is directly traceable to Darwinism and the rejection of the early chapters of Genesis."
e) "Christians believe that the majority of the fossils are the remains of creatures that died in the flood."
f) "Self-replicating molecules first appeared on the earth 3.5 billion years ago."
g) "Life has proved entirely explainable in physical-chemical terms."
h) "To doubt evolution is to doubt science, and science is only another name for truth."
i) "The cradle of every new science is surrounded by extinguished theologians."

Do you agree with me that all these sentences are false?

Why are they wrong? Most are exaggerations, and you can detect the wishful thinking behind them. Some, like (a.), are sheer guesswork. Some sound witty but are simply wrong. The last statement (i.) is particularly unfortunate: it was made by T H Huxley towards the end of the nineteenth century, at about the time when the *monk* Gregor Mendel was doing the experiments that would found the science of genetics.

Chapter 10

The Two Books

At the start of *The Origin of Species*, even before the Title Page, Darwin quotes three authors. Here is the third quotation:
"To conclude, therefore, let no man out of a weak conceit of sobriety, or an ill-applied moderation, think or maintain, that a man can search too far or be too well studied in the book of God's word, or in the book of God's works: divinity or philosophy; but rather let men endeavour an endless progress or proficience in both."
Bacon: *Advancement of Learning*

I wonder if you were surprised to read such words at the beginning of *The Origin of Species*? Darwin was a thoughtful man and he would not have put them in carelessly. Bacon wrote of "two books", the Bible (the book of God's word) and Nature (the book of God's works) and it is clear that he expected his readers to treat them with equal dignity and to learn as much as possible about both.

This idea of the Two Books was not just Bacon's view. In his book *Rocks of Ages* Stephen Jay Gould refers to Newton, Halley, Boyle, Hooke, Ray and Burnet, all devout men who set the foundations of modern science in late-seventeenth-century Britain. These scientists argued that God would not allow any contradiction between his *words* (recorded in scripture) and his *works* (the natural world)." [1] If Bacon could come back amongst us today he would be dismayed to discover that there were naturalists who held the book of God's word in contempt and some believers who claimed that scientists were telling lies about the book of God's works.

Underlying the Great Debate is a clash of creation stories, representing two different worldviews. Until *The Origin of Species* was published the accepted belief in the

West was that plants, animals and human beings had been created by God, as described in Genesis. Darwin's book introduced a new creation story. Never mind that in Darwin's day it was often theologians who accepted the new story and scientists who rejected it: the genie had been released from the bottle. Hence the Great Debate. People are puzzled why the debate has not long since been settled, but the events and stories refer to the distant past, and neither side can *prove* that their view is the right one.

When I first heard of evolution I only knew of two positions in the debate: those who accepted it and those who did not. The first group rejected the Genesis story as myth or poetry. The second group believed in a literal interpretation of Genesis, and rejected evolution as a human error (or even as a lie of the Devil!). There must have been, between these two extremes, people who accepted the Bible but did not suppose that Genesis was science, and so did not think that it clashed with evolution: perhaps God created simple living things and then allowed them to evolve into the forms we have today. This idea today is called "theistic evolution". I heard that phrase for the first time half-way through my teaching career, when I attended a Creationist Event in the local town. The speaker was a creationist author and speaker with a doctorate in chemistry. He attacked evolution in general and the idea of millions-of-years-of-evolution in particular. Much of his talk was on radiometric dating which, he argued, was completely unreliable. He was more fiercely against theistic evolutionists (TEs) than atheist Darwinists. I now know that this is not unusual. Most of the creationist authors that I have read take a dim view of TEs, believing that they let the side down, by taking too light a view of Scripture and, by accepting evolution, making God responsible for millions of years of waste, cruelty, disease and death.

The spectrum of views

Today there is a wide spectrum of positions in the Great Debate. Here is a rough list (rough, because the categories overlap and it becomes untidy in the middle):

Ultra-Darwinists: neo-Darwinists who are outspokenly atheist. Sometimes called Darwinian fundamentalists.

Darwinists: the default position. NDS adherents: probably agnostics, and some atheists but you don't know because they don't go on about it.

Other evolutionists: those who accept evolution but who are critical of the NDS and looking for alternative mechanisms for evolution.

Theistic evolutionists (TEs): Those who believe that God used evolution to create living things. The majority of TEs accept the Darwinian mechanism of evolution. A minority are looking for alternatives. [2]

Adherents of Intelligent Design (IDers): Opponents of the NDS who support a sophisticated form of the Argument from Design based on modern studies, particularly biochemistry. They usually do not appeal to Genesis and "have no position" on the age of the earth.

Old-earth creationists (OECs): Biblical literalists who accept the evidence that the earth is millions of years old.

Young-earth creationists (YECs): Biblical literalists who believe that the earth is only thousands of years old

Within these groups there are subgroups. There are "adaptationists" and "Gouldians" amongst the Darwinists, who argue about minor details of how natural selection works. Some theistic evolutionists accept the NDS; others do not. Some TEs are liberal in their theology and ready to explain away large chunks of Scripture as myth or folk-tale, while others are "Bible-believing" but don't think that Genesis is a scientific account. There are a number of OE creationist positions, each with a different way of reconciling the biblical record with geology. There are distinct groups of YE creationists, each with a different emphasis. Non-Darwinists are accused of being antiscientific, but there are holders of science doctorates and professorships in every one of the above groups.

We are used to it now, but isn't this an odd situation? You don't get a spectrum of views on mechanics or chemistry. This situation is more like the proliferation of political parties, or church denominations, than anything else in science. (This leads to the obvious retort: "Ah, but has this *really* much to do with science?")

The extreme ends of the spectrum

In Part I there was little mention of the most argumentative and high-profile participants in the debate, those at the extremes, because I wanted to write about the *science* of evolution. It is hard to think of an ultra-Darwinist as a scientist when he repeatedly tells the world that he is an atheist. Atheism is not a *scientific* position. A scientist is someone who has an open mind: an atheist is someone with a mind which is shut and bolted on at least one topic.

Creationists and ID supporters were not included in Part I either, not because their techniques in the laboratory or the field are unscientific, but because they go outside science for pieces of missing data and hold beliefs about biology which are not scientifically testable. (Note, I am not saying they are wrong to do this; just that it is not *scientific*.) Whether or not it is true that the world has an "appearance of age" or that there is an Intelligent Designer, these matters are not part of science.

Science, by definition, deals with the natural world: "Explanations that cannot be based upon empirical evidence are not part of science." Therefore, if an author states that *biology* shows us that there is no afterlife or God-given morality, he has stopped being scientific. If another author tells us that God brought young dinosaurs to the ark built by Noah, he has also stopped being scientific. Science cannot study these things nor pronounce on them.

A general reader might wonder, "Why is *science* so all-important?" Answer: it is not; it is a technique and a tool of thought. But it is continually being dragged into the debate. It sounds much grander to say "Your views are an attack on science!" than to say "Your views are an attack on my views." Ultra-Darwinists complain that Intelligent Design is an attack on science. In a recent book we were told that ID endangers the global dominance of the USA in science and therefore presents "the gravest of threats to the American economy, which is driven by advances in science and in the technology derived therefrom." [3] This is an extraordinary statement. Are we really to believe that a non-mainstream theory about a historical aspect of biology is going to affect the American oil industry, and Walmart, and NASA?

Creationists claim that what *they* are offering is science. There is a Creation Science Movement, an Institute for Creation Science and a Creation Research Society. Both sides claim that they are the true scientists, because of the huge prestige of science. But the prestige attaches to *hard* science, because of the advances in technology and medicine over the last two centuries, not to historical sciences such as evolution and creation science. Science itself is not at stake, for two reasons. Firstly, you cannot actually attack *science* any more than you can attack mathematics or logic; you can have poor science or wrong calculations or fallacious reasoning, but those are not "attacks" on the actual disciplines themselves. Secondly there are numerous scientists with doctorates, or professorships, on both sides of the debate. The last geology lecture I heard was given by a research geologist with a doctorate, currently working in the Grand Canyon. He is a YE creationist and believes in the 6-Day creation as described in Genesis; he is a first rate geologist and knows far more about rocks and fossils than I do. Any disagreements between him and some of his listeners were not about the facts, but how they were interpreted.

Those who occupy positions at the extreme ends of the spectrum have several things in common: they refer to their opposite numbers as "fundamentalists" and see the Great Debate as a black-and-white matter. Both believe in mechanisms (fiat creation on the one hand and natural selection on the other) which can explain *anything at all*; and both see their opponent's views as not merely wrong but preposterous. Two other characteristics are these:

Blinkered views: narrowness and reductionism

Biblical narrowness consists in the claim that the early chapters of Genesis must be taken literally or "as they were intended to be understood"; that there is no room for any wider or richer interpretation. This narrowness is strongly defended and strongly attacked. It is defended by the claim that "The Way is narrow", that the Scriptures are inerrant, and that anything else is compromise. It is ridiculed by the accusation that such views are obscurantist.

Scientific narrowness takes the form of reductionism. Reductionism is "the belief that the higher levels of integration of a complex system can be fully explained through a knowledge of the smallest components." [4] It is the idea that that you can explain bodies completely in terms of cells; cells in terms of cell organelles, the latter in terms of molecules . . and so on down. From the point of view of research, a reductionist approach makes good sense: you study something by taking it apart. But you must not then say that the bigger thing is *nothing but* a colony or a compound of the smaller ones. The Niagara Falls consists of more than just a lot of water, and water itself is more than just atoms of oxygen and hydrogen.

A simple analogy of the pitfalls of reductionism is this: suppose we wanted to analyse a passage from Shakespeare. We want to find out why people think *Romeo and Juliet*, say, is a great work. We can analyse the scenes, then the speeches, then the sentences and then the actual words. Next we can study the spelling and the actual letters of the words. Finally we can analyse the chemical composition of the ink and the paper. Anyone can see that somewhere along the way we have lost the plot (in both senses).

Reductionism involves a lack of imagination. Truth is grasped by the imagination as well as the reasoning intellect. The great scientists were men of insight and imagination. Think of Leonardo da Vinci, Newton, Faraday, Einstein and Darwin himself. They were broad-minded and open-minded, as well as logical. Reductionist reasoning gets narrower and narrower until you end up in a cul-de-sac. One thinks of G K Chesterton's remark in *Orthodoxy* that "the madman is not the man who has lost his reason. The madman is the man who has lost everything except his reason." In the same chapter Chesterton makes this remark about materialism:

"As an explanation of the world, materialism has a sort of insane simplicity. It has just the quality of the madman's argument; we have at once the sense of it covering everything and the sense of it leaving everything out." [5]

This is supremely true of reductionism; explanations end up with "So what?"

Reductionism in evolutionary biology is particularly unfortunate. In all the talk of genes and macromolecules and replication, the "elephant in the room" - the fact that the cell is *alive* - is conveniently forgotten. Nucleic acids cannot replicate on their own. Dead cells cannot do anything. If you "reduce" life too far you simply stop the process you are studying. Reductionism turns living beings into passive objects.

It is a retreat from common sense. You end up by stating that human beings are *merely* machines for propagating DNA, or robots, or something equally silly. You are left with a sort of world-weary negativism. This is how Joy Davidman described her philosophy when she was still an atheist:

"Life is only an electrochemical reaction. Love, art and altruism are only sex. The universe is only matter. Matter is only energy. I forget what I said energy is only." [6]

Reductionist biology is also retrogressive, according to Augros and Stanciu: whereas modern physics is breaking free from its old mechanistic programme, biology - with its machine model of life, and its attempt to reduce consciousness to physiology - is moving in the opposite direction: back towards an older, discarded physics. [7]

Just after I wrote this section I discovered and read James Le Fanu's book *Why Us?* and came across this paragraph:

"It is not the least of the ironies of the New Genetics and the Decade of the Brain that they have vindicated the two main impulses to religious belief - the non-material reality of the human soul and the beauty and diversity of the living world - while confounding the principle tenets of materialism: that Darwin's 'reason for everything' explains the natural world and our origins, and that life can be 'reduced' to the chemical genes, the mind to the physical brain." [8]

Intolerance

Those who occupy territory at the extreme ends of the spectrum are intolerant of other views. This is worst at the ultra-Darwinist end. YE creationists, it is true, will not budge one inch on the literal inerrancy of Scripture, and view all evolutionary views as hopelessly compromised. But they do not try to stifle debate. Some ultra-Darwinists, on the other hand are dogmatic. Anyone who questions evolution is mad or wicked and should not get a hearing. Magazines should not accept creationist letters. Schools must not mention creationism in class. Parents should not be allowed to teach creation to their children: in some cases ultra-Darwinists have extended this to any religious beliefs. Who would have expected that the Galileo episode would be re-enacted with scientists playing the part of the Inquisition?

You can always tell that an ideology is thoroughly bad (and insecure, for that matter) when it denies freedom of thought, speech or movement to its opponents, whether it is a political system that builds a Berlin Wall to keep its people in, or a religion that threatens its adherents with death if they convert to a different faith. A scientific ideology is now falling into the same trap. [9]

Returning to Bacon's 'Two Books', it is clear that they must, ultimately, be found to tell the same story. This conviction will not be shared by ultra-Darwinists who have no time for the Bible, nor by those creationists who have a dim view of science. The resolution of the debate will not be promoted by the polarisation of these two groups, who are - in the words of a recent article – "feeding off each other." The atheists speak of their anger or sadness when some creationist initiative comes into the news, but they enjoy the opportunity to jeer. And the creationists relish the quarrel. In 2002 there was a public controversy over the teaching of creation science in a school in the North of England. [10] Newspapers, radio and television all carried interviews and comments. The dispute even reached the House of Commons and the school was mentioned - not unfavourably - by the then Prime Minister, Tony Blair. However most of the response was critical, even hostile. Several well-known evolutionists warned of the dire consequences if pupils were told about any alternative to evolution, and demanded that the school should be re-inspected (it had received a glowing Ofsted report). A friend invited me to a creationist meeting a few days later, and the organiser began by referring to the dispute about the school and added, rubbing his hands with glee, "These are exciting times to be a creationist!"

Here is an curious fact: in this dispute about the teaching of creation science in the school, six bishops joined forces with several atheists to warn the Prime

Minister about the "danger of creationism." The Bishop of Oxford claimed on Radio 4's "Thought for the Day" that creationism brought Christianity into disrepute. At the same time a group of nearly thirty scientists and academics signed a letter to the Education Secretary supporting the stance of the school, pointing out that they wanted schools to teach children how to think - not what to think. Isn't this intriguing? It parallels the situation just after the publication of Darwin's *Origin of Species* in which many churchmen accepted evolution and many scientists rejected it.

Notes to Chapter 10

1. S J Gould, 2001, *Rocks of Ages - Science and Religion in the Fullness of Life* Chapter 1

2. Francis Collins remarks that "theistic evolution" is a terrible name. He has suggested the term "BioLogos"; Denis Alexander uses "evolutionary creationist" as an alternative to TE.

3. From *Intelligent Thought*, 2006, edited by John Brockman (Introduction). This book consists of 16 essays attacking the Intelligent design movement.

4. This definition is from the glossary to *What Evolution Is* (2001) by Ernst Mayr.

5. *Orthodoxy* (1908, paperback 1961), G K Chesterton (Chapter 2).

6. See the Foreword to *Smoke on the Mountain* (1955) by Joy Davidman.

7. *The New Biology* (1987), by Augros and Stanciu

8. James Le Fanu *Why Us?* Chapter 10

An excellent recent account of the pitfalls of reductionism can be found in *God's Undertaker* (2007), by John Lennox, in which he describes three different types. See Chapter 3 "Reduction, reduction, reduction . . ."

9. Captain Acworth, back in the 1950s, told me of a letter received from his friend, C S Lewis, after he had read some of the EPM literature. Lewis wrote "I must confess that it has shaken me: not in my belief in evolution, which was of the vaguest and intermittent kind, but in my belief that the question was wholly unimportant . . . What inclines me now to think you may be right in regarding it as the central and radical lie in the whole web of falsehood that now governs our lives is not so much good arguments against it as the fanatical and twisted attitudes of its defenders."

10. Emmanuel College, Gateshead.

Chapter 11

Christian parts of the spectrum

The spectrum of views on evolution is cumbersome because science and religion are both involved, something that Stephen Jay Gould disapproved of. In his writing he championed the concept of Non-Overlapping Magisteria (NOMA).[1] A *magisterium* is a domain of authoritative teaching, such as science or religion. According to Gould, science deals with the facts of the natural world, and religion with human values and purposes. He claimed that these domains do not overlap; therefore science and religion can never be unified and any conflict between them is false.

Creationists and IDers do not accept NOMA. Gould summarised the NOMA principle briefly in an attack on Phillip Johnson in the *Scientific American* of July 1992 stating that science treats factual reality, while religion is concerned with human morality. Johnson replied that the power to define "factual reality" is the power to govern the mind, and thus to confine "religion" inside a naturalistic box. God's commandments cannot provide a basis for morality unless God actually exists. The commandments of an imaginary deity are just human laws dressed up as divine law. Johnson accused Gould of putting forward a view (naturalistic metaphysics) which "relegates both morality and God to the realm outside of scientific knowledge, where only subjective belief is to be found."[2]

NOMA is not accepted by ultra-Darwinists either (they have no time for *any* religious belief) so it is not clear how many people go along with it. Some of the problems of the Great Debate are caused by people muddling up the two magisteria by trying to give scientific answers to religious questions and *vice versa*; but there are questions to which the answers do overlap, such as "Does evolution or creation best explain the genesis of nature?" and "Which came first, mind or matter?" It is because of this overlap that the spectrum of views is untidy.

Old-earth Creationism

Members of the Evolution Protest Movement were OE creationists. Their pamphlets and books attacked both the theory of evolution and the belief that natural selection was its mechanism. They deplored the social and political effects of Darwinism, reflected in the dictatorships of Hitler and Stalin. They were indignant that glory that should have gone to God went instead to Evolution - and, indirectly, to High Priests of Evolution like Darwin and the Huxleys.

The EPM published many pamphlets and booklets. There was little mention of the age of the earth, or reference to Genesis beyond its first three chapters. Instead they emphasised the *moral* aspect of evolutionism, and condemned teachers and the BBC for promulgating evolutionary ideas.

Until the rise of Young-earth creationism (after the publication of *The Genesis Flood* in 1969) the EPM represented the views of Christians who did not accept evolution. There was no consensus amongst OECs about how the early chapters of Genesis relate in detail to the historical sciences. They agreed that Creation was miraculous (sudden, rather than by a long process), but not about how the biblical timing relates to geological dating. Some OECs claimed that the "days of creation" were geological ages, a view taken by a number of the EPM leaders. Others believed that each "day" represents a long period of time on the basis of a verse in the New Testament: " . . With the Lord a day is like a thousand years, and a thousand years are like a day." (2 Peter 3:8). Yet others accepted some version of the "Gap Theory": the idea that the geological ages fit in between two portions of the Biblical record. [3]

At least two other methods have been put forward by which the six days of Genesis can be reconciled with the vast age of the earth:

One idea put forward was that the days were *Days of Revelation*. God revealed to Moses the details of His creation over the course of six days. This idea ties in with the impression that we are watching a drama unfold as Genesis 1 proceeds. The curtain rises on a featureless, dark stage (the earth without form and void), there is a cry of "Lights!" and the grand play starts.

Another idea was called the *Framework theory*, which proposes that the division of the Creation account into six days is a literary device, in the same way that a piece of music might be divided into six movements. [4]

OE creationism did not generate the disdain that the YE creationists elicit from evolutionists. OECs accept the scientific evidence for the age of the earth and try to reconcile it with the biblical account; they also avoid the problems of "Flood Geology." They meet the evolutionists half-way, so there is less to disagree about. However they could be seen to inhabit the worst of all worlds, being neither orthodox in their biological views nor snugly within the biblical literalist camp.

The EPM was renamed the Creation Science Movement in 1980, which now embraces YE creationism.

Young-earth Creationism

This is what is usually meant by "creationism" today. The rise in numbers and influence of YE creationists in the last quarter of a century can be traced to two things: the impact of *The Genesis Flood*, and the hostility and aggressive atheism of the ultra-Darwinists.

Young-earth creationism is based on a literal interpretation of the early chapters of Genesis: a reasonable position because that's how Scripture has been read for many centuries. So YECs believe that the various "kinds" of plant and animal were created separately over the space of six days, and that Adam was the first human being, linked through recorded genealogies to the historical figures of the Old and New Testaments. YECs also believe the following, which, though not explicitly stated in Scripture, have been argued using Scriptural principles: the earth is only a few thousand years old, the "days" of creation were 24-hour days, there was no animal death before Adam's Fall and most fossils are the remains of creatures that died during Noah's flood.

The age of the earth is around 10,000 years.

In 1650 Archbishop Ussher [5] worked out the exact date of creation by adding up the ages of the people listed in the biblical genealogies: he arrived at the figure of 4004 BC. That was a reasonable suggestion for the mid-seventeenth century, but later discoveries in archaeology and geology made that date untenable for most people. Throughout the nineteenth century the date of creation was pushed further back in time as new discoveries were made: today most scientists believe that the earth is some billions of years old. YECs disagree; they accept that there are gaps in the biblical genealogies, but reckon that Ussher wasn't that far out. They do not accept the "millions-of-years" age of the earth for these reasons:

Firstly, absolute ages for rocks are based on radiometric dating (the rate of decay of radioisotopes into daughter elements, such as potassium into argon or uranium into lead). These measurements are not of *time*, but of the relative *amounts* of parent and daughter elements in a given rock. YECs claim that this method is unreliable because it is an extrapolation based on assumptions: that the rate of decay is constant, that the starting proportions of the elements are known, and that there has been no leakage or inwards diffusion of the elements over millions of years.

Secondly, some rocks known to be formed in historical times (for example lava flows that have formed beneath the sea) have been erroneously given dates of millions of years. If what can be checked is wrong how can you trust what cannot be checked?

Thirdly, these dates are manipulated in order to allow time for evolution to take place. When different methods, or different readings, give a wide range of ages for a rock, the oldest will be chosen.

YE creationists believe that there are positive reasons to believe the earth is young. Most YEC books include a list of evidences for a young earth, based on measurements of the amount of salt in the sea, or dust on the moon, or the height of coral reefs. (Note that these measurements depend just as much on extrapolation and preliminary assumptions as do the radiometric measurements indicating a vast age for the earth. [6]) Finally, YECs claim that many natural phenomena give an "appearance of age"; i.e. that they were created in final or adult form (see "Gossery", Chapter 13). If so, it removes the point of making the measurements.

Creation took place during the course of six 24-hour days.

The disagreements between OECs and YECs about the meaning of the word "day" are often linguistic arguments about the right translation of the Hebrew word y*om*, which can mean "'day" or "period of time" depending on the context. This discussion is not fruitful because we can never finally *know* what an ancient writer meant by the word he used, and therefore there is no casting vote about whose interpretation is right.

There was no animal death before Adam's Fall.

The Fall of Man is described in Genesis 3, followed by God's Curse. YECs claim that before these tragic events there was no disease, parasitism or animal death in nature. Verses to support this claim include Genesis 1:30 (animals were all herbivorous) and Genesis 1:31 (when Creation was complete, God pronounced it "Very Good"). Thus, animal death is the result of the Curse. This belief is crucial: the literature rails against the "millions of years of death, disease, and pain before Adam's sin" which contradict the clear teaching of Scripture and are thus unacceptable to biblical Christians.

YECs will not accept those "millions of years" at all. When I discussed this with a creationist friend I discovered that the matter was not negotiable. This is the argument against any form of theistic evolution: God is good and evolution is cruel and horrible; God would *never* have used it to create animals and plants.

Nearly all fossils are the remains of creatures that died during Noah's flood.

This follows directly from the last point. Fossils are the remains of dead creatures, and death resulted from the Fall; therefore fossils must have been formed *after* Adam appeared on the earth. This applies to all of them, including the dinosaurs. There are millions of fossils, found in many hundreds of feet of sedimentary rock: they must all have died during an overwhelming catastrophe, namely, the global Flood recorded in Genesis, Chapters 6 to 9. Apart from things like mollusc shells and fish bones which accumulate at the bottom of the sea, most fossils are formed when animals have died in abnormal circumstances, such as sand storms or floods. And a global flood is not impossible; after all, 70% of the world is still flooded today, and even the land above water is covered with huge areas of sedimentary rock, most of it laid down under water. Some fossils have formed in more recent times, but the vast majority of fossils result from Noah's Flood.

Critics of the YEC position all agree that this makes nonsense of standard geology. Ingenuity has been shown by YEC scientists in finding ways round this, but key questions are left unanswered. Where did all the material come from? How could thousands of feet of rock of different types (including those formed from wind-blown sands, slowly-accumulated chalk, or volcanic debris), and which have been eroded, buckled, faulted and riddled with volcanic seams, all form in the single year of the flood? The whole of the geological column condensed into a few months?

YECs (including working geologists) have come up with some Just So stories, but have made no impact on the wider scientific world. Geologists have studied rock formations for over two centuries; they have reached conclusions based not just on a few measurements and observations but on masses of *interlocking* data. Flood geologists have made some good points, about fossil graveyards, and rapid sedimentation and so on, but have not succeeded in persuading ordinary scientists that standard geology is wrong. You will not convince an evolutionist that he is wrong by telling him that the defence-mechanism of the bombardier beetle disproves evolution, nor will you convince a working geologist that he is wrong by telling him that pollen grains have been found in Precambrian deposits. Science isn't like that.

To non-YECs, flood geology seems desperately contrived, and when I chat to my creationist friends I am reminded of T E Huxley's comment to Darwin about gradualism: "you have loaded yourself with an unnecessary difficulty."

The claim that fossils result from Noah's Flood is a distinctive feature of YE creationism. It did not appear in the EPM literature and is not part of ID. For YECs, as we have seen, flood geology follows from the belief that animal death is the result of the Fall. But is it? We must go back to source: to the actual facts in biology, and to actual verses in Scripture. The Fall and the resulting curse are in Genesis Chapter 3, and if you read the passage you will see that animal death is not mentioned. The belief that animal death resulted from the Fall is an *interpretation* of the two verses in Genesis 1 mentioned above and of certain New Testament passages. Christian critics of YE creationism do not disregard Scripture, but challenge this interpretation.

A world without animal death may be imaginable - just - but it is not the world we know. Firstly, animals have life-cycles. If the adults never died, the living space for that creature would be filled up, and reproduction would stop. There would be no more babies, calves, cygnets, tadpoles or caterpillars. Secondly, nature's ecosystems are based on food-chains and the more complex food-webs. A world without animal death would be a world with just two-member food-chains (plant eaten by herbivore: grass eaten by rabbit). There would be no ladybirds, ground beetles or spiders; no kingfishers or hawks; no weasels, cats or tigers. These considerations are not arguments against the YEC position, but comments on the implications of that position.

YECs - and kind-hearted people in general - regard animal death as something ugly and tragic, perhaps because we usually think of those animals which are close to us in size or intelligence, such as apes or dogs. Is the death of a mouse or a fish tragic? What is the alternative? The idea of an *immortal* grasshopper or frog is meaningless, even comic. Perhaps it isn't death itself that is abhorrent, but the thought of violent, painful and disfiguring death. Creationists argue against evolution on the grounds that "God would never have used such a cruel and wasteful" method of creating. But it is not evolution that is cruel and wasteful (if those are useful words to use) but nature itself. There is a huge mortality in nature whether evolution happens or not. If evolution is non-Darwinian, the charge of "cruel and wasteful" has even less force. A more cogent criticism of Darwinian evolution is that it is the embodiment of *selfishness*, as summarised in the phrase "the survival of the fittest." Selfishness is the mainspring of human wickedness. It is true that animals have no moral sense, but there is a case to be made that God would not use selfishness as a means of creation.

One of the key verses used by YECs is Genesis 1:31 "God saw all that he had made, and it was very good." Surely, a perfect creation would have no carnivory in it. But "very good" does not mean "perfect". It is hard to think straight about this. Someone walking in beautiful countryside might describe it as "very good", but all countryside, however attractive to look at, contains decay and parasitism and sudden death. A lot depends on our mood. One moment we agree with Robert Browning that "God's in his heaven - All's right with the world !" and the next we are appalled by the apparent cruelty of nature. Some Victorians were happy to accept carnivory or parasitism as examples of the provision and ingenuity of the Creator. But Darwin himself was horrified by parasitism (he cited the larvae of an ichneumon wasp feeding inside a still living caterpillar) and he referred to "the clumsy, wasteful, blundering low & horribly cruel works of nature!" [7]

Our preoccupation with cruelty and suffering is a modern one. Our forefathers were tougher-minded than we, and readier to accept suffering as part of life. Animal death is not described as evil in the Bible, and there is no commandment against eating meat. Dislike of carnivory might be a real moral advance; but it might reflect the squeamishness of over-civilised people who have lost touch with their hunting or farming roots, and who are frightened of pain. C S Lewis pointed out in *The Problem of Pain* that the great religions in the world were "first preached and long practised in a world without chloroform." The existence of suffering is used as an argument against the existence of God. This is a modern idea, and one most likely to be found amongst comfortably civilised people.

Intelligent Design

Intelligent Design differs from YE creationism: it is not specifically Christian, does not depend on a literal interpretation of Genesis and takes no particular view of the age of the earth. Some who promote Intelligent Design ("IDers") accept evolution.

IDers claim that the Darwinian mechanism for macroevolution, as well as lacking empirical evidence, is inadequate as an explanation of the diversity, complexity and purpose in the living world. They claim that the evidence indicates an *Intelligence* behind nature.

The history of this movement has been chronicled by Thomas Woodward in *Doubts about Darwin*. Woodward argues that the movement began with the publication of Denton's *Evolution: a Theory in Crisis* [8] which "cleared a path" for ID. Another landmark event was the publication of *Darwin on Trial* by Phillip Johnson which attacked Darwinism along four lines of argument:

i. Biological evidence, especially that from palaeontology, falsifies the Darwinian story.
ii. Macroevolution is based on the philosophy of naturalism, not on empirical evidence.
iii. Darwinism is supported by poor argument and faulty logic.
iv. Darwinism is the central cosmological myth of modern culture, and hence quasi-religious, not scientific. [9]

Johnson is not putting forward an alternative mechanism for evolution or a detailed explanation for the diversity of living things. He is arguing against the acceptance of Darwinism as a scientific certainty, when it is a worldview, or even a religion.

Johnson is taken more seriously by the academic world than YE creationists, partly because of his position as Law Professor at Berkeley and partly because he involves fellow-academics in conferences. He has been heavily criticised by evolutionists, who mostly miss the point. One criticism is that he is not a scientist (he is a judge and professor of Law). But the fact that he is an "outsider" is in his favour. He raises the *philosophical* problems with Darwinism and Naturalism, and a judge is well placed to do that. Johnson deals with the logic of the arguments, the weight of the evidence behind a particular claim and the question of whether a case has been made or not. Rather than being *for* Intelligent Design his book is *against* neo-Darwinism. It elicited a damning 4-side review from Stephen Jay Gould in *Scientific American*; criticism which probably did no harm to Johnson, and actually confirmed his main point. Gould simply didn't get what Johnson's book was about. All his criticisms of Johnson's views were from within the neo-Darwinian paradigm, but Johnson was asking whether the paradigm was itself adequate and whether it should be changed. To a philosophical Naturalist, there is nothing in existence outside the universe discoverable by science, and so philosophy and religion and God *must* all be found within his scheme. For him, Naturalism is the biggest "Russian doll", and all the other dolls, like philosophy or religion, must fit inside it. But Johnson argues that there is no evidence from biology that Naturalism is the biggest doll; he makes the case that there is a bigger one still. Gould's criticisms did not weaken, but strengthened, Johnson's case.

Johnson provided no scientific replacement for neo-Darwinism, but it is reasonable for a judge to say what did *not* happen even if he doesn't yet know what did. It clears the ground for others who can then proceed with new investigations. Whether or not ID is accepted, Johnson's criticisms of Darwinism are cogent.

Another ID landmark was the publication of *Darwin's Black Box* by Michael Behe in 1996. Behe is a "scientific insider", a research biochemist, who put forward the idea of *irreducible complexity*. This describes any system (a machine, or cellular mechanism) which depends on every part being present and working simultaneously; and which cannot work at all if any part is missing. He gave the illustration of a simple mouse-trap of five parts: if any is missing the trap doesn't work. Such a structure could not have been assembled piecemeal; it has to be the product of design. Behe maintained that biochemists have discovered many such systems in the body (e.g. the blood-clotting mechanism) and at the level of the cell (e.g. intracellular transport devices or the bacterial flagellum). Such irreducibly complex systems could not have evolved by natural selection: they wouldn't work at all unless every component was present and these could not all have evolved at once. They are evidence of *design*. ID is partly negative, listing all the defects of the NDS, and partly positive. Paley was right after all; just as a watch has been put together by a purposeful and intelligent person, so living things are the products of a purposeful intelligence.

IDers do not disallow evolution: some, including Behe himself, accept the common descent of living things. Nor do they necessarily identify "the Designer" as God; they mostly leave that question open and are happy that Design is "on the table", alongside other theories. This caution disappoints YECs.

Intelligent Design is disdained by evolutionists who claim that it is a "God-of-the-gaps" position: scientific explanations are accepted until they run out, then a Designer is invoked. Of course, this is not a good criticism if there really *are* gaps - which is what IDers claim. Evolutionists also claim that the idea of irreducible complexity is flawed. A cellular structure or biochemical pathway may *appear* to be irreducibly complex, but new research may show that it's not. Already several of Behe's biochemical examples have been challenged.

Darwinists also criticise IDers for holding a view which is untestable, and which therefore acts as a brake on science. IDers reply that criticisms of untestability (or "unfalsifiability") apply equally to natural selection as the cause of a past event. As for ID acting as a brake on science, Behe has replied that the purpose of science is to discover how things work and where they came from: it is "the search for truth, not just merely the search for materialistic explanations". [10]

IDers are accused of producing few peer-reviewed articles for the science press. This is a strong criticism in normal science, but the situation is changed if a paradigm shift is proposed. Most scientists are trained in the same way; sharing the same ideas, hearing the same lectures, using the same textbooks; anyone exploring controversial ways of thinking will find it very difficult to get his work published. Failure to win acceptance or find a publisher does not necessarily mean that the work is unscientific. When a researcher questions neoDarwinism, even from within its own ranks, he draws heavy fire from the evolutionary establishment. It is not permitted to doubt any aspect of evolution because it is now a worldview, not just a part of biology.

Darwinists regard ID as a bigger threat than creationism. They have traditionally ridiculed creationists, referring to them as "red-necked" or "Bible-bashers" but this name-calling is inappropriate for the more academically secure IDers, particularly as they discuss information theory and biochemistry instead of Flood Geology. Instead of jeering at a soft target, Darwinists marshal serious arguments against ID. A good example is the book entitled *Intelligent Thought* edited by John Brockman: sixteen essays attacking Intelligent Design by fairly well-known and high-powered authors who find it threatening.

Theistic Evolution

The previous three positions are fairly well-defined, but TEs do not form a homogeneous group; nor any sort of 'party'. A "theistic evolutionist" is someone who believes in God and happens to accept evolution. He or she may rarely think of the two together. TE is probably the default position of countless people, but it would be hard to find out. Some give TE a different name (Chapter 10, Note 2). Many of the books by TEs are actually about something else, with evolution mentioned in passing. Only one of the six sections in *God and the Biologist* by Professor Berry is about evolution. *One World* by John Polkinghorne, about the interaction of science and theology, has just one page on evolution and half a dozen sentences elsewhere. In the same author's *The Way the World is* - also about the relationship between Science and Christianity - the word "evolution" does not appear in the index, and "C Darwin" gets just one sentence.

A key concept in TE is the idea of *complementarity*. A person might say in one context that the earth is held in its orbit by gravitational forces, and in another, that God sustains the world. Both explanations are true but in different ways: they are complementary. A theistic evolutionist believes both that God is the Creator and that evolution has happened. A concise description of complementarity is given by Professor Berry:

"It is wholly consistent with both science and Scripture to insist that God is the Creator, but also that he worked through mechanisms which we may discover through scientific research." [11]

Complementarity can be illustrated from the world outside science. Suppose someone asks, "Why did Ophelia die?" You could reply, "because she went mad and drowned herself", or "because Shakespeare wrote her death into Act IV of *Hamlet* for dramatic reasons." Both are true - they are complementary answers.

TEs insist that science and the Bible provide answers to different questions, and in different ways. Scientists try to answer When ? and How ? and theologians try to answer Why ? and Who ? Most theistic evolutionists accept Darwinism (the NDS).

Theistic Evolution is abhorred by creationists. They accuse TEs of being compromisers, who want to hold on to their religious beliefs but retain respect in scientific circles: half-hearted Christians attempting to fit the Bible into an ungodly worldview. They attack the very phrase "theistic evolution" as a contradiction in terms. Behe claims that TEs are kidding themselves if they think that theism is compatible with Darwinism, because God presumably set nature up to ensure a particular outcome, but Darwinism relies on randomness. [12]

Ultra-Darwinists are double-minded about TEs. They welcome them as allies against the creationists, but ridicule the God of cruelty and waste that the TE position implies.

Theistic evolutionists accuse IDers of relying on "God of the gaps" arguments. These ultimately fail, because if "the designer" is used as an explanation to cover a gap "he" will shrink as knowledge increases. A lot depends on the nature of the gap. It is not wise to put God in any gap in *scientific* knowledge. That is what John Polkinghorne was referring to when he wrote:

"The one God who is well and truly dead is the God of the Gaps. His job was to pop up as the explanation, so-called, of what otherwise could not be understood . . .

. . . If God is God he is to be found everywhere, not just in the murkier corners of the world he has made." [13]

John Lennox has made two extremely important points about this. Firstly, the argument cuts both ways:

"It is very easy to say 'evolution did it' when one has not got the faintest idea how, or has simply cobbled up a speculative just-so story with no evidential basis. Indeed, as we have seen, a materialist *has* to say that natural processes were solely responsible, since, in his or her book, there is no admissible alternative. As a result it is just as easy to end up with an 'evolution of the gaps' as with a 'God of the gaps'. One might even say that it is easier to end up with an 'evolution of the gaps' than a 'God of the gaps' since the former suggestion is likely to attract far less criticism than the latter." [14]

Secondly, science cannot fill all gaps in knowledge, nor can it answer every question that human beings ask. John Lennox distinguishes between "bad gaps" and "good gaps":

"As an example of a bad gap, we might think of Newton's suggestion that God occasionally had to tweak some of the orbits of the planets to bring them into line. That kind of gap we would expect to be closed by science because it falls within the explanatory power of science to settle. Good gaps will be revealed by science as not being within its explanatory power. They will be those (few) places where science as such points beyond itself to explanations that are not within its purview." [15]

Just as creationists are dismayed by TEs, the latter disapprove of creationists, whom they believe are misguided. Firstly, there are better ways of resisting attacks on religious belief than just maintaining what atheists write off as obscurantism. Secondly, teaching creationism to children is setting them up for big problems at school or later in life. Most TEs are equally critical of the fundamentalists at the other end of the spectrum, the ultra-Darwinists. In *The Selfless Gene* Charles Foster takes both Dawkins and the creationists to task for claiming that the theory of evolution by natural selection excludes the existence of God, and he deplores the teaching of both. [16]

Notes to Chapter 11

1. In *Rocks of Ages* (2001) Gould had over 200 pages on the relationship between science and religion: the idea of NOMA is at the heart of it. He also described NOMA in a short essay: *Leonardo's Mountain of Clams and the Diet of Worms* (1998)

2. This exchange of views is described in the Epilogue to the second edition of *Darwin on Trial* (1993) by Phillip Johnson.

3. The best-known "gap" is that between verses 1 and 2 of Genesis 1. For the actual words see my Ch.1, Note 3

4. The Framework theory is put forward by Roger Forster and Paul Marston in their book *Reason and Faith*. Those who accept it may not be creationists. A theistic

evolutionist might well take the first chapter of *Genesis* to be a literary device.

5. The date of 4004 BC comes from *Annals of the Ancient and New Testaments* published in 1650 by Archbishop Ussher of Armagh

6. *What About Origins* (1978) Chapter 5 Monty White has a table giving a list of 76 different ages for the earth (ranging from a few hundred to millions of years !) based on different processes such as influx of elements into the ocean via rivers or decay of the earth's magnetic field.

7. From a letter to J Hooker, 13 July 1856

8. *Evolution: a Theory in Crisis* (1985) by Michael Denton. Woodward also refers to *The Mystery of Life's Origin* (1984) by Charles Thaxton, Walter Bradley, Roger Olson and Dean Kenyon

9. *Darwin on Trial* (1993) by Phillip Johnson.

10. From a conversation with Lee Strobel (see *The Case for a Creator* 2004)

11. *God and the Biologist* (1996) ; Chapter 3, by R J Berry

12. In *The Edge of Evolution* (2007) by Michael Behe

13. From *One World* (1986) Chapter 4, by John Polkinghorne

14. *God's Undertaker*, (2007) Chapter 9, by John Lennox

15. ibid, Chapter 11

16. *The Selfless gene* (2009) by Charles Foster (Ch. 2: "A Tale of Two Cities and Two Bigotries"); the two cities are

North Oxford (home of Richard Dawkins) and North Kentucky (home of the Creation Museum).

Chapter 12

Fossils

" . . if evolution means the gradual change of one kind of organism into another kind, the outstanding characteristic of the fossil record is the absence of evidence for evolution."

Phillip Johnson; *Darwin on Trial*

Fossils provide the only *hard* evidence (in both senses of the word 'hard') for the occurrence and nature of the plants and animals that lived in the remote past. Fossils are physically present: you can examine them, measure them and record the exact place you found them.

Why do fossils exist? You do not often see a dead animal lying on the ground. Nearly all are eaten by carnivores or scavengers and their bones (if they are vertebrates) are scattered and gradually disintegrate. Dead plants - or parts of plants, like leaves - fall to the ground and are destroyed by insects, worms, fungi and bacteria. You would have to go to unusual places, like deserts or mineral springs, or a dying lake with hardly any bottom-dwellers, to find any fossilisation happening today. Fossils are usually formed when there is quick burial following sudden catastrophes, such as floods or mud flows or clouds of volcanic ash or sand-storms: and catastrophes are infrequent.

The chances of a particular fossil being found are remote. First the organism had to be buried quickly. Then parts of it, together with whatever material buried it, had to harden and become rock. That rock had to remain undamaged; over countless millennia it might be sheared, baked or shattered, any of which would have destroyed the fossil. Finally the part of the rock containing the fossil would have to become accessible, and the rock weathered just the right amount for the fossil to appear at the surface.

Too little weathering and the fossil is hidden, still embedded in the rock: a few more years and the fossil itself has weathered away. And in the relatively short space of time when the fossil is at the surface and still undamaged it has to be found by someone who knows it to be a fossil, and who collects it. There are billions of fossils in the world, but only a tiny minority are ever discovered and examined. It is remarkable that rare fossils (such as *Archaeopteryx*) were ever found.

There is disagreement amongst experts about the importance of fossils as evidence for evolution. Professor Grassé, the French zoologist, insisted that the process of evolution is *only* revealed through fossils; that only palaeontology can show the course of evolution: everything else is just hypothesis [1]

Another zoologist, Mark Ridley, disagreed, going so far as to write that "no real evolutionist" relies on the fossil record as evidence in favour of the theory of evolution as opposed to special creation. In his view, arguments from "observed evolution" (microevolution), from biogeography, and from the classification were much more important." [2] (That was over a quarter of a century ago. Today writers would include developments in microbiology such as gene sequencing.)

The fossil evidence does few favours to either side in the Great Debate. This "hard" evidence has not clinched the argument. Like the other evidences for evolution, it can be interpreted in different ways, much to the exasperation of both sides.

Fossils and the NDS

Critics of neo-Darwinism claim that the fossil record does not support it:

In the *Origin of Species* there is a chapter headed "On the Imperfection of the Geological Record" in which Darwin claimed that "the number of intermediate varieties

which have formerly existed [must] be truly enormous."
He goes on:

"Why then is not every geological formation and every stratum full of such intermediate links? Geology assuredly does not reveal any such finely-graduated organic chain; and this, perhaps, is the most obvious and serious objection which can be urged against this theory. The explanation lies, as I believe, in the extreme imperfection of the geological record." [3]

Modern critics point out, with suitable quotations from palaeontologists, that not much has changed in the last 150 years. Although thousands more fossils have been discovered, the "finely-graduated organic chain" has never been found.

Most fossils do not show gradual changes. When I was studying geology at Cambridge we had to become familiar with over five hundred invertebrate fossils (ammonites, sea urchins, and so on), to identify them and know which rock they were found in. This would have been impossible if they had all graded into each other. We found that it was just as easy to separate fossils into their genera and species as it is for living animals. Each fossil species was unique, and did not change over thousands or even millions of years. There were a few exceptions to this, such as the group of Cretaceous sea urchins which formed a developing series, but the changes were so small that they only amounted to microevolution. These fossil facts do not support Darwinism.

The NDS explanations for the Cambrian Explosion (sometimes called ""Biology's Big Bang") are unconvincing. There is little that is provisional or intermediate about the Cambrian organisms: they are already members of most of the large invertebrate phyla when they first appear. The NDS describes species splitting and forming genera which diverge further into separate families, then orders, and so on. But the Cambrian explosion and subsequent evolution is the wrong way round with the major differences in form and body plan appearing first and the minor differences later.

Quoting the saying, "The Past is another country", creationists argue that the fact that organisms differ as you go from stratum to stratum is no more evidence for evolution than the fact that organisms differ as you go from country to country across the world. The fauna of South America and Africa are very different, after all. One may speculate about curious modern animals like the duck-billed platypus or penguins. If they were only known as fossils, would they have been hailed as missing links? If modern dogs were only known as fossils how many different genera, or even families, would they be put into? No one could guess that they all belonged to the same species. Fossil remains can only tell you a limited amount about the actual animal or plant when it was alive.

Fossils and Young Earth Creationism

Critics claim that the fossil record does not support the creationist view either. Many years ago J B S Haldane pointed out that to deny evolution you would have to affirm that "species were wiped out and their successors created on a slightly fantastic scale." [4] He put his finger on a salient point. Most fossils are of different animals and plants from those alive today. There are countless animals and plants in the fossil record which have no living representatives, including over 2000 different ammonites, 17,000 different trilobites and getting on for 700 different dinosaurs. It has been estimated that 95% of fossils are of extinct forms. YECs claim that most fossils were formed at the time of Noah's flood. If that is so, why should the animals and plants that survived the flood be different from the ones that died out? In the biblical account (*Genesis* chapters 6 - 9) the animals that survived were representatives of all the ones that drowned. Ingenious attempts have been made to explain that climate change or earth movements can account for the extinction of the dinosaurs after the flood, but what about the countless species of mammals that have died out? To account for the extinction of thousands of species after the flood, and the distribution of the surviving organisms, YE creationists invoke large scale volcanic activity, earth movements, climate change and even a form of continental drift. Not one of these things is recorded in the Bible.

There are other problems with Flood Geology. Firstly, here is part of a letter from a friend who is a theistic evolutionist:" . . it claims that the fossils were all laid down in the flood; but a very conservative estimate of the number of vertebrates that must have been living at the time of the flood in order to produce the number of fossils we see gives each animal a plot the size of a hearthrug - not enough to feed a rabbit, let alone a cow!"

Secondly, the order of vertebrate and land plant fossils which you find as you progress up through the rock systems reflects the classification. (Not just an evolutionary classification, *any* classification.) There are fish at the bottom of the series; amphibia, reptiles and non-flowering plants towards the middle; mammals and flowering plants at the top. No explanation so far given by flood geologists for this sorting has convinced their critics. Haldane put the case succinctly when he stated that "a single skeleton of a mammal in Silurian rock would wreck the theory of Evolution." [5]

Such a fossil would be strong evidence for flood geology, but none has ever been found.

Thirdly, and overtopping all, is the *time* problem. Critics of flood geology will never accept that most of the major sedimentary rock formations in the world were laid down in less than a year. If flood geologists are right then the whole of historical geology is wrong.

Some general points

If neo-Darwinian evolution is correct, what would you expect to see in the fossil record? Not the Cambrian explosion, or stasis. You would expect, as Darwin did, gradual change and countless transitional organisms. And you don't get that. What would you expect to see if flood geology is right? Not trilobites and dinosaurs, but a higgledy-piggledy mixing of the remains of dead creatures; the same ones as those that survived the flood. And you don't get that either.

For a non-specialist it is hard to obtain the facts. Creationist authors claim that there are no fossils representing "missing links". They state that there are no intermediate organisms between between fish and amphibia, between amphibia and reptiles, or reptiles and mammals. But main-line books on vertebrate palaeontology describe intermediate fossils between all of these groups. *Eusthenopteron* is placed between fish and amphibia, *Seymouria* between amphibia and reptiles, *Thrinaxodon* or *Diarthrognathus* between reptiles and mammals. When challenged with the opposing group's claims the creationists always say that the intermediate organisms are not really intermediate, and the evolutionists always say "Oh, yes they are" and both sides complain that the other's interpretation is biased by their world-view.

The words "intermediate" and "transitional" are easily confused. *Intermediate* is a trivial word referring to an item's place in a group or list. A van is intermediate between a car and a lorry. In biology you could describe a sea lion as being intermediate between an otter and a dolphin, without it implying anything about their relationship. *Transitional* refers to a much closer link: a transitional organism is one that is on the line of ancestry between two forms. If you find three different species of sea urchins A, B, and C showing progressive development as you move up the Chalk strata, it is reasonable to claim that B is not just an intermediate between A and C but is transitional between them, meaning that A gave rise to B and that B gave rise to C.

These words are misused. In a recent article in *New Scientist* the author wrote that Darwin predicted that transitional fossils would be discovered, and that "millions . . have been uncovered." This is misleading not only because the reader is unsure whether "millions" refers to individuals or species, but because it is *intermediate*, not *transitional* fossils which have been found.

The argument between creationists and evolutionists is often just about these words. Both sides accept that there are intermediate organisms (that's common sense). But creationists deny that there are any transitional organisms, certainly between the bigger groups such as classes or phyla. Unfortunately you simply cannot prove it, either way; a point made clearly by the palaeontologist Colin Patterson who reminded us that although fossils can give us a lot of information, "one thing they can never disclose is whether they were ancestors of anything else." [6]

The disagreement between the two sides is clear when any fossil series is examined. Take that of the horse fossils from the Cenozoic, which are described in almost every text-book of biology. Neo-Darwinists regard the series as incontrovertible evidence for evolution. Arranged in sequence, as in the textbooks (and in our display at Millfield), they show gradual changes from genus to genus, the height of the animal increasing, the number of toes decreasing, the molars gradually changing from simple 'browsing' teeth to complex 'grazing' teeth, and so on. But opponents of evolution claim that this series is just a human construction. Old World and New World fossils are mixed up together in order to tell the story. The first three forms (*Eohippus*, *Orohippus*, and *Epihippus* actually decline in size, and anyway, the size range of all the creatures is the same as that of modern horses (tiny Falabellas to huge Shirehorses). The traditional sequence has *Eohippus* followed by *Orohippus* (both Eocene) at the bottom, *Pliohippus* (Pliocene) in the middle and *Equus* (Upper Pliocene through to recent) at the top. But *Eohippus* and *Equus* have 18 pairs of ribs, *Orohippus* had 15 pairs, and *Pliohippus* had 19 pairs. Why should the number of ribs vary as you go up the sequence?

Evolutionists continue to see the horse series as good evidence for evolution *in spite of* this sort of criticism. These discrepancies are exactly the sort of thing you'd expect in an actual living system of multiple branching lines of animals over many millions of years. And opponents of evolution *always* resist this conclusion and claim that there are only a few horse-like genera (perhaps just three: four-toed, three-toed and single-toed,) which did not evolve one from another.

Another focus of controversy is the evolution of the whale. Creationists claim that there are no intermediate fossils between whales and land animals, but books on evolution list a series of such intermediates. [7] Once again, the two sides simply attack each other's arguments. Creationists point out that *Pakicetus* is only known by its skull, for goodness' sake, and that *Ambulocetus*, otherwise a very telling intermediate, has several key parts of its skeleton missing.

All "missing links" are important to the two sides in the dispute. But those nearest-to-the-knuckle are the ones in the line leading to human beings. It is hard to be sure of the data, because everyone involved in the arguments has a powerful vested interest. One book tells us that all the hominid remains in the world could fit into a suitcase. Another states that there are "enough human remains to fill a graveyard." Human Evolution historically is a subject spoilt by frauds and rumours of fraud, hidden specimens and damaging disclosures, misidentifications and imaginary reconstructions. Creationists gleefully describe a whole series of bad mistakes from the past: Java Man, Piltdown Man, Nebraska Man, Pekin Man - all dutifully discovered with fanfares, described in detail, illustrated in magazines and reconstructed in museums but subsequently discovered to be imaginary, fraudulent or of dubious authenticity.

Today evolutionists point to the series of fossils - *Australopithecus africanus*, *A. afarensis*, *A. (*or *Homo) habilis*, *Homo erectus*, *H. sapiens* - and claim that there are no missing links on the journey from ape to man. (These fossils do not form a straight line of ancestry but are seen as progressively-more-human twigs on a bush of species.) They claim that this evidence is about as good as it gets with fossils. Creationists reply that, no matter how complete the series, you simply have a group of apes (*Australopicethus*) on the one hand and humans (*Homo*) on the other. They argue that all the humans are members of the same species (*Homo sapiens*) and that the range of fossil forms is no wider than the range of modern races. Exasperated evolutionists complain that creationists will *never* accept any intermediate organism as a missing link, so what's the point of arguing? Creationists reply that evolutionists fudge the data: if a fossil turns up in the wrong place (in the time sequence or geographically) they simply reclassify it or ignore it. All that matters is evolution: fossil data are accepted if they support evolution and discarded if they don't.

A famous fossil missing link is *Archaeopteryx*, a Jurassic bird from the fine-grained lithographic limestone of Germany. It was discovered just after Darwin's *Origin* was published. It is a true bird but with many reptilian characters, and therefore a good example of an "intermediate" fossil. Unfortunately it is the wrong sort of intermediate. It has *complete* wings and feathers, not half-way ones. It is a mosaic of avian and reptilian characters, so it does not answer the questions that evolutionists need to ask: how do you get from a front leg to a wing? where do feathers come from? Sir Fred Hoyle found this fossil unsatisfactory as a link because of its isolation, "with no steps in the record from reptiles to *Archaeopteryx* or from *Archaeopteryx* to birds, as the Darwinian theory requires." [8] Hoyle was not a creationist, but they would agree. They insist that *Archaeopteryx* is a perfectly good bird which happened to have claws on its wings (like the living bird called the hoatzin) and teeth in its beak (like several other fossil birds).

It is easy to see why the great debate continues. There are millions of fossils: real, hard facts from the past. The two sides of the debate will never agree on how the fossil record should be interpreted. If all the facts and "knockdown arguments" were on one side, the dispute would have evaporated a century ago, but they are not. If all the creationists were uneducated and the evolutionists clever scientists (or vice versa) the discussion would also have petered out by now. But they are not; and it has not. The debate today is as vigorous as ever.

Summary: "Something is gravely wrong"

In their book Augros and Stanciu stated that much of the fossil record contradicts the premises and corollaries of Darwinian theory. They support this claim by citing leading palaeontologists all of whom have demonstrated that the fossil record is jerky and not gradual.

They comment: "Something is gravely wrong with a theory that forces us to deny or ignore the data of an entire science." [9]

YE creationism insists on some form of "flood geology." This tries to show that the geological column is a fabrication, that nearly all the world's sedimentary rocks were laid down in a few months, less than ten thousand years ago, during which billions of dead animals and plants were sorted into classification order.

This invites the same comment: "Something is gravely wrong with a theory that forces us to deny or ignore the data of an entire science."

Notes to Chapter 12

1. *L'Evolution du Vivant* (1973) by Pierre Grassé (English translation: *Evolution of Living Organisms*, 1977) Grassé was a distinguished French zoologist who was an evolutionist but did not accept the NDS.

2. *Who Doubts Evolution?* (1981) an article in *New Scientist* of June 25 by Mark Ridley

3. *The Origin of Species* (6th Edition), Chapter X

4. *Fact and Faith* 1934 Chapter 1 by JBS Haldane

5. ibid, Intro vi

6. *Evolution* 1978, Chapter 11, by Colin Patterson

7. In the collection *Dinosaur in a Haystack* (1996), Stephen Jay Gould devotes Essay 28 ("Hooking Leviathan by Its Past") to the question of whale evolution. It starts with the embarrassing tale of Darwin's bear (Darwin invented a most unlikely Just So story about bears becoming whales); but the whole essay is an attack on creationists and ends with a gloat at their expense.

8. *The Intelligent Universe* (1983) by Fred Hoyle

9. *The New Biology* (1987) by Robert Augros and George Stanciu. The palaeontologists cited are Steven Stanley, David Raup and Niles Eldredge.

Chapter 13

Gossery

Philip Henry Gosse was an eminent Victorian naturalist. In 1857 he published a strange book called *Omphalos*, with the subtitle *An Attempt to Untie the Geological Knot*. I had heard both good and bad things of this remarkable - and now very rare - book, and I managed to borrow a copy. Gosse was an unusual man. He was an outstanding naturalist, a Fellow of the Royal Society and an acquaintance of Hooker and Darwin. He has been called the Father of Marine Biology, and Stephen Jay Gould referred to him as "the David Attenborough of his day, Britain's finest popular narrator of nature's fascination." [1] Gosse's enthusiasm for natural history is reflected in the note he jotted down in his diary on the day his only son was born: "E. delivered of a son. Received green swallow from Jamaica." [2]

But Gosse was also a member of the Plymouth Brethren and an enthusiastic upholder of Scripture, with an insistence on a Creation Week of six 24-hour days. He wrote *Omphalos* in an attempt to reconcile the narrative from the Bible with what he knew of geology and natural history. His idea was actually quite simple, so simple that he need not have written a long book; a pamphlet would have done just as well. In his introduction he directs the reader to just two pages in the book, in which he sets out his two main points. The rest of the book consists of endless examples of his idea, wordily written up and accompanied by beautiful woodcuts of animals and plants, but really just padding.

His first point is this: all living things have life-cycles (he called them "circles") and any act of creation would have to break into a life-cycle somewhere. Therefore just after something has been created it will give the *appearance of age*. A newly-created horse will have teeth which appear to be worn down as a result of years of grazing. Trees, though newly created, will look as if they have been growing for decades; inside each trunk will be many years' worth of annual rings. And, the supreme example, Adam will have a navel even though he had never been born (*Omphalos* is the Greek for navel). Gosse named such illusory characters "prochronic" as opposed to structures formed after Creation which he called "diachronic".

Gosse's second idea was very strange. He argued that the history of the whole world could also be considered a part of a circle into which creation must break. Therefore the rock strata, the fossils and the light from the distant stars are also prochronic. The fossils are real in the sense that you can handle them and study them, but they never actually existed as living organisms! In this way Gosse tried to reconcile the first chapters of Genesis with the great geological ages: the latter are just an appearance, not real.

Omphalos is extraordinary because there are hundreds of pages of good geology and biology providing examples of structures with prochronic histories that never actually took place! Unfortunately for Gosse, this multiplication of examples, instead of making the reader more convinced, makes the idea seem more silly. It is interesting to compare this with *The Origin of Species*, which Darwin published only two years later, and in which the multiplication of examples adds to his arguments.

Criticisms of *Omphalos*

Gosse's work is almost completely forgotten today but, it has been referred to in essays by two excellent authors, Stephen Jay Gould and Dorothy L Sayers who come to almost opposite conclusions about it.

In *Adam's Navel*, Gould referred to *Omphalos* as "spectacular nonsense" and "reason at its most perfectly and preciously ridiculous" and it is easy to see why.

Firstly, the language is quaint and old-fashioned, with "Lo!" and "Ha!" and "Let us go yonder . . ." (another contrast with *The Origin of Species*. Compared with Gosse, Darwin writes in a sober, scientific way: if you didn't know, you'd think the two books were half a century apart rather than two years). Secondly, the "world life cycle" is an extraordinary idea, without any supporting evidence, and the analogy with life cycles of organisms doesn't work. The claim that fossils are real but that the animals of which they are the remains never actually lived is preposterous. Gosse's contemporaries certainly thought so: the book was a failure, and poor Gosse was devastated. His hopes had been so high, but were cruelly dashed. His son, Edmund Gosse, many years later wrote:

"This was to be the universal panacea; this the system of intellectual therapeutics which could not but heal all the maladies of the age. But, alas! atheists and Christians alike looked at it and laughed, and threw it away." [3]

Gould's main criticism was not about the detail but the principle: there is no way of finding out whether *Omphalos* is right or wrong. Gosse's idea is completely untestable, and therefore it is not part of science.

The other author is Dorothy Sayers (writing many years earlier than Gould). In a thought-provoking piece called *Creative Mind* [4] she describes Gosse's idea (let us call it *Gossery*) but without mentioning him by name. She points out that if theologians had chosen to think of God as a great imaginative Artist instead of a Great Engineer they could have offered an unusual interpretation of the facts of Nature "with rather entertaining consequences." They could have seriously advanced the theory that God had at some particular moment created the universe complete with all the vestiges of an imaginary past. Well, that is what Gosse had tried to say, of course, but Sayers makes an additional comment. Gosse's idea is an "extravagant assumption"; if you think of God as a mechanician, but if you think of him as working in the same way as a creative artist, it becomes natural. "*It is the way every novel in the world is written."* In a novel the characters will live a particular story within the pages of the book, but there will be many allusions to their pasts; i.e. what happened before the opening of the first chapter. These allusions are literary "fossils." Sayers asks us to imagine one of the characters looking at the evidence of his own past: she claims that there are no data, nor any imaginable line of reasoning, that he could use to prove whether or not that past had ever actually happened. After developing the idea she concluded that, if theologians had chosen to take up that position, the result would have been entertaining. Not only that, it would have been a very strong position, because it couldn't be upset by scientific proof.

In this passage Dorothy Sayers is playing with an idea rather than putting forward a serious proposition. But isn't it interesting that two clear, analytical minds can come to two such different conclusions about Gosse's idea? Gould tells us that theories that cannot be tested in principle are not part of science; that we should reject *Omphalos* as useless rather than wrong. Sayers says of the same idea that it would have been a powerful one because it is cannot be upset by scientific proof.

They agree on one thing. Whether Gossery is a useless idea (Gould) or a powerful one (Sayers), both are saying that it is outside science.

Why do I mention Gosse at all? Well, partly because, like Gould and Sayers, I find the whole idea curious and fascinating. But there are other reasons:

Appearance of Age

Gosse put his finger on a very important point. Any sudden creation must involve some "appearance of age". A literal reading of Genesis chapter 1 must involve Gossery, for every newly-created thing, plant or animal or star, would appear to have a history. Such a reading may be right, but it cannot be *scientific* because there is no way of testing "before" and "after".

Creationist authors are well aware that sudden creation implies Gossery. In *What About Origins?* A J Monty White wrote that when God created the heavens and the earth, the sea and everything in them, he did so giving them a *superficial appearance of age*. The plants and animals were created fully mature (" the fish had never hatched from eggs . . " and so on) [5] A critic would object that though this might be reasonable biblical interpretation, it comes oddly at the end of a chapter in which scientific methods are used to demonstrate that the earth is only a few thousand years old. Surely it doesn't much matter *how* old the earth appears, if the appearance is, anyway, superficial. As Alan Hayward has pointed out, if God made the world look old at creation, there is no use hunting for evidence of youthfulness. [6] Some authors have invoked Gossery to explain how distant stars could be millions of light years away and yet created only a few thousand years ago: the light was created already on the way here. But this is just an assertion of which Gould might have remarked, as he did of *Omphalos*, that we cannot devise a way to find out whether it is either right or wrong. It is an untestable notion.

Questions raised by Gossery

Whether Gosse's notion is right or wrong, it does draw attention to some questions that might otherwise not be asked. For Believers:

Is creation involving the appearance of age an example of miracle? Is it wise for creationists to use the term *Creation Science* for the study of miraculous events? Can God act only by miracle? Are miracles always instantaneous?

and for everyone:

How far can you insist upon *testability* for past events? Is macroevolution by natural selection any more testable? Is testability by the scientific method the only form of testability that matters?

Mechanic or Artist

There is much food for thought in the point mentioned by Dorothy Sayers: is there any good reason why we should think of God as Creator in mechanistic rather than artistic terms?

Last-Thursdayism

This follows on from Gossery. As there is no possible way to distinguish between prochronic (apparent) and diachronic (real) events, the world *could* have been created in 4004 BC with a completely imaginary pre-history. But in that case the line could be drawn anywhere at all. Hence the intriguing idea of "Last-Thursdayism" which I am told is on the Internet. The idea is that the whole universe was created last Thursday, complete with all its prochronic evidence, and your and my prochronic memories.

That is a joke, but it has a serious side to it. There is a sense in which the world did begin, for each of us, on the day of our birth. Everything that happened before that date is like the prologue to the novel or play in which we are a character. And this world will come to an end, as far as each of us is concerned, on the day that we die. Gosse's scheme may or may not apply to the history of the world of nature, but something of it applies to each one of us.

If you read *Mind at the End of its Tether,* the last, terribly sad, book by H G Wells, you find the converse of what I wrote. Wells was coming to the end of his life (he was 79) and believed that the world was coming to the end of *its* life.

"The end of everything we call life is close at hand . . . (p.1)

"The writer is convinced that there is no way out or round or through the impasse. It is the end. (p.4)

"events . . go on and on to an impenetrable mystery, into a voiceless limitless darkness . . .the door closes upon us for evermore. There is no way out or round or through. (p.13)

"Our doomed formicary is helpless as the implacable Antagonist kicks or tramples our world to pieces. Endure it or evade it; the end will be the same . . " (p.17) [7]

Many of the passages in *Mind at the End of its Tether* are about evolution ("This idiot's tale . . . comprehends the whole three thousand million years of Organic Evolution.") and when Wells writes about chlorophyll, crinoids and notochords he becomes quite animated. But he soon gets back to his theme of pointlessness and cruelty and despair. He had long ago turned his back on what he called "religion", and can find no comfort at all in philosophy or science. The short book ends in bleak darkness. [8]

Notes on Chapter 13

1. See the essay 'Adam's Navel' which is Chapter six in the collection *The Flamingo's Smile* (1985) by Stephen Jay Gould.

2. Recounted in *Father and Son* (1907), the autobiography of Gosse's son Edmund, written some years after Phillip Henry Gosse's death (he died in 1888). Edmund's book is unreliable in some details, but is fine literature and of great interest.

3. *Father and Son*, Chapter V

4. *Creative Mind* was originally an address given in 1942. It is printed in the collection *Unpopular Opinions* (Gollancz) of 1946.

5. *What About Origins?* (1978), Chapter 5, "Dating Processes" by Dr. A J Monty White

6. *Creation and Evolution* (1985), Chapter 8

7. *Mind at the End of its Tether* (1945) by H G Wells

8. Compare this passage from Bertrand Russell: "There is darkness without and when I die there will be darkness within. There is no splendour, no vastness anywhere, only triviality for a moment and then nothing." *Mysticism and Logic* (1918) (quoted in Chapter 9, "The Silence" of James Le Fanu's book *Why Us?*)

Chapter 14

The Book of God's Word

A YE creationist recently said to me, "I see that you're still sitting on the fence." His remark provided food for thought. Glib responses came to mind: "As there are at least seven positions on creation and evolution there must be at least six fences" and "If the data on both sides of a question are insufficient to settle it, the fence is the best place to sit." But the point is that the creationists at one end of the spectrum and the ultra-Darwinists at the other *want* there to be just two positions, either "For the Bible, against evolution" or "For evolution, against the Bible." (Some Darwinists try to enlarge this dispute by using the word "science" instead of evolution; denouncing creationism as an attack on science. This is nonsense: many creationists are themselves scientists.) This 'either-or' debate is mistaken for two reasons: it is based on a misunderstanding of what the Bible is, and "evolution" is taken to mean "Darwinian evolution" which is not the same thing.

Appropriate Understanding

Those who take the Bible seriously agree that it is an extraordinary book - but then often read it in an ordinary way. One of its remarkable properties is that it carries multiple meanings. All life is there, light and darkness, wisdom and folly, but also richness and complexity. It may be compared to an intricately embroidered Oriental carpet, woven from thousands of coloured threads, on which a pessimist can follow the pattern of the black threads and an optimist can follow the pattern of the golden threads. Ten different people will gain ten different things from a Scripture reading. Over the years, *one* person may read different things from the same passage. Often you hear someone say, "I've been familiar with this chapter all my life and never seen that verse before!" or "*Now* - for the very first time - I know what those words mean."

One result of this multiplicity of meanings is that people select different parts of the Scriptures to emphasise. (This is one reason why there are so many denominations.) Critics sneer at discrepancies and apparent contradictions in the Bible. The discrepancies are not a problem. Anything reported by several different witnesses will contain discrepancies, and the Bible has many authors. The writers of the Gospels wrote many decades after the events took place so its not surprising if some of the details don't agree. (If they *did* agree, if all the stories and statements dovetailed together neatly, the critics would have devised a conspiracy theory.) As for the contradictions, I am reminded of something that happened during my first few months as a schoolmaster. The Headmaster called me in to discuss a pupil who was doing badly. I told the Head that I thought the boy lacked gumption. "Yes", he replied, "I wish he'd go and break a window!" For a moment I was startled because the Head was a strict disciplinarian. But I knew what he meant about that particular boy. This was not double-mindedness on the part of the Headmaster, just wisdom; an apparent contradiction, not a real one.

Anything complicated and living will be a mass of apparent contradictions, exceptions and discrepancies. Nature is full of them. Mammals give birth to their young - except the platypus and echidna which lay big, shelled eggs like those of a reptile. Fish breathe with gills - apart from those which have lungs. The yew is a conifer, but has no cones. One European weevil exists only as females. Most orchid flowers are actually upside down . . . . . .

Of course the Bible has discrepancies and contradictions in it! It is not some text-book, a codified course for living. It multifarious and untidy, like life itself. Much of Jesus' teaching is the same. The apparent contradictions are because he was addressing different people, with different needs, at different times. CS Lewis has pointed out that Jesus's teachings cannot be reduced to a system: 'He utters maxims which, like popular proverbs, if rigorously taken, may seem to contradict one another. His teaching therefore cannot be grasped by the intellect alone, cannot be "got up" as if it was a "subject."' The attempt to pin him down is "like trying to bottle a sunbeam." [1]

The Old Testament has diverse writings with many sublime passages and also some strange stories. It is bitty and multifarious - the Bible says so itself: the writer of the book of Hebrews starts his letter: "When in former times God spoke to our forefathers, he spoke in fragmentary and varied fashion through the prophets. . . " Preachers tell us that God meets us where we are; and we are all in different places. Therefore, when we read the first chapters of Genesis, those that concern us in the Great Debate, different interpretations can be true for different readers. A child or an adult; a farmer or a biologist, will see different aspects of the truth. The words are true for a particular person at a particular moment, as long as that person is taking the words seriously. This could be called the principle of Appropriate Understanding.

Every teacher knows that you must adjust what you teach to the age or capacity of your pupils. In the Junior School you might allow children to think that plant "feeding" has to do with fertilisers. For GCSE you would give a simple explanation of photosynthesis. But at A Level you would teach about absorption spectra and the Calvin cycle. It isn't that what you teach at A Level is "true" and the rest is "false" but that you handle the facts according to the level of understanding of your pupils.

This is not an exact analogy to understanding Genesis. Firstly, what happens in photosynthesis is a matter of present-day facts and repeatable experiments, not a narrative from thousands of years ago. Secondly, although you meet different levels of chemical complexity when you study photosynthesis you don't change the subject;

you are still studying where plants get their materials and energy from. Bible passages, however, have different meanings in different categories. Thirdly, someone's interpretation of Genesis may have little to do with that person's age or intelligence.

"Appropriate understanding" is not relativism. A relativist claims that there is no absolute truth, that all ideas are valid, that - in some lazy way - this is true for me and that is true for you; let's just agree to differ. No; the truth is there, but it is multifaceted and hard to discover: people move towards it at their own speed, and understand it in their own way. So we must give other people space and not insist that our interpretation is the only right one.

People change their minds. I know of several people, including scientists, who accepted evolution for years but for whom a literal reading of the biblical account made more sense when they made a detailed study. They are now creationists. [2] I know of others who started off as creationists but later became convinced that the evidence supported evolution. A friend of mine changed his mind *twice* (he is now a convinced theistic evolutionist). He is a clever mathematician and a good scientist. When he was a creationist he could marshal excellent arguments for his views, and he will now give you good arguments to show that creationism is wrong. He knows by experience that neither view is silly.

Appropriate Understanding partly depends on appropriate wording. Are you preaching, discussing geology or telling a story to children? You would use different words for each. It would be a silly person who sat through a service of Nine Lessons and Carols muttering to himself: "that is not scientific." The Great Debate becomes tiresome when the words of Scripture, or science, are taken out of context: straw men are set up and attacked, and no one gets anywhere.

Finally, the principle of Appropriate Understanding only applies to biblical *stories* where wise people have come to no consensus about the interpretation. It is not suggested that everyone should have their own "take" on the Ten Commandments!

Seriously, not literally

Christians refer to the Bible as the word of God. [3] By this they do not mean - obviously - that God wrote it. They believe that it was written by ordinary human beings under the inspiration of God. The 66 books have many different authors who wrote in many different styles. Some books, like the Book of Psalms or the Song of Solomon, are poetry in almost our modern sense. Both are packed with poetic images ("let the rivers clap their hands, let the mountains sing together . . ") that no one would dream of taking literally. Other books, such as I and II Kings, or the four Gospels, are historical in the sense that they describe day to day matters in ordinary language. Most of the books in the New Testament are letters, some of them only a page or two long. Considering that they were written nearly two thousand years ago, they are astonishingly fresh and relevant: more so than many books written between then and now.

But many of the books of the Old Testament (and Revelation in the New Testament) cannot be classified in the same way. "History", "poetry" or "science" are modern categories: they are not appropriate for these ancient writings. Two descriptive terms are missing from our discussions about Scripture:

1. Heightened Language

A term is needed for the non-everyday language of much of the Old Testament. Huge chunks of it (about 40%) are said to be "written as poetry." Unfortunately the word *poetry* is not a useful one. For many people poetry is a form of highbrow intellectual entertainment, rather like ballet. For others, poetry is make-believe, as opposed to history or science. If you tell a creationist that the first chapters of *Genesis* are poetry he replies, "No, they are true!"

In the absence of a better word I will call it "heightened language". This language, with its figures of speech, is used by many of the biblical authors, particularly the Prophets. Here are two examples from Isaiah:

"The people walking in darkness have seen a great light"

"those who hope in the Lord will renew their strength. They will soar on wings like eagles; they will run and not grow weary . . "

and one from Joel:

"I will repay you for the years the locusts have eaten..."
For people familiar with the Bible, it is easy to forget how very different this heightened language is from other writing. We have got used to it. The idiom, and many of the actual phrases, of this ancient form of writing have become ingrained in Western people because we are the end-product of centuries of Christian culture.

The word "symbolic" is appropriate for some parts of the Bible, particularly the Book of Revelation, but not for most of the Creation story. A symbol is something chosen to represent a quality or a concept. A lion is a symbol of strength; Uncle Sam is the symbol of the American people. But the fruit-trees and the birds and the livestock in the Creation story are simply themselves; they are not symbols of something else. (Possible exceptions are the two named Trees in Genesis 2 and the serpent in *Genesis* 3.)

Heightened language carries more weight than everyday words; it speaks to our imagination, not just our intellect. Two mistakes can be made with it. One is to take it literally; the other is not to take it seriously. Genesis is a hard book to interpret because it is not clear in some passages whether the language is heightened or not.

2. Vertical History

The other term missing from our vocabulary is one which describes biblical happenings in which God is a participant. "History" is the wrong word, just as "poetry" is the wrong word for the heightened language of the Bible. History is the study of past *human* affairs: it is about pharaohs and kings and battles: it is earthbound or "horizontal." In many books of the Old Testament, and in Revelation, the history is "vertical", with descriptions of what God is saying and doing. The Introduction to Genesis in the *Good News* Bible states: "Throughout the book the main character is God, who judges and punishes those who do wrong, leads and helps his people, and shapes their history."

Note that it is what happens to the men and women which is "history." There is no word for the total event, including God's participation. Even the human history in the Bible leans towards what matters *spiritually*, often omitting details that an ordinary historian would include. For example, in the account of the birth of Moses, we are told the names of the two midwives, but not that of the reigning pharaoh. [4]

One author writes that it is essential, when reading a passage of Scripture, to "decide whether it was historical, or scientific or poetic before attempting to interpret it." This advice is wrong because these are modern categories; also, we are separated from these writings by thousands of years, thousands of miles and huge differences in knowledge and culture. It is impossible to be sure that our interpretation is right.

The book called Genesis

The matter is complicated because Genesis consists of three very different narratives of unequal length. The first is just forty-eight sentences long (Gen. 1^1 to 2^3.) It is unique: 'vertical history' told in simple, direct words, many of just one syllable, equally understandable to a child or a sage. Nothing is said about *how* God created. The account is not written as poetry; but because of the greatness of the events recorded it has the *effect* of poetry: one is left in awe.

The second narrative is also short, just a couple of chapters (Gen. 2^4 to 3^{24}). Unlike that of Chapter 1, which describes global and universal events, this very different second part is homely and local: the four named rivers puts Eden firmly in the Middle East. The events recorded here (whatever name you care to give them - 'fantastic', 'supernatural' or what) - are outside normal human experience. Some details, such as the "Flashing Sword" are opaque to the imagination.

The third narrative forms the whole of the rest of the book (Gen. chapters 4 to 50). The story starts off as vertical history, but as chapter follows chapter, it becomes more and more 'historical' in the modern sense. In the early chapters, God is described almost as a participant in the human life of these early people: he has conversations with them ("The Lord then said to Noah . . . "). In the later chapters the writer speaks of God infrequently; in the last ten chapters he is mentioned just once, when he "spoke to Israel in a vision at night" (chapter 46)

If "history" is the incorrect word to describe the earlier chapters, what can we use instead ? The obvious word is "story", but it may be misunderstood. A YEC friend was dismayed when I called the Genesis creation account a story because for him a story was something invented, therefore untrue. If you speak of a story, sooner or later you will hear it called "a *mere* story." I heard someone say, "Why cannot you just accept Genesis as history? God said it, I believe it, and that settles it." I would reply: I *do* accept the account in Genesis and do not believe that people who take it as history are wrong. Those, such as children and the very wise (and a lot of people in between) who are able to take the story as it stands are to be envied. But many people do not believe that the story is history, and it doesn't help to contradict them. The story is fruitful to those who take it seriously, whether or not they take it literally. We must give other people space: it is not necessarily wrong to take a passage of Scripture literally, but it is important to allow other people not to do so. Also, people discover things at their own speed. [5]

The authors of several modern books state that the important thing is to take the Genesis account "as the author intended" rather than "literally". I agree; but I think that the author intended them as story. And they make a splendid story:

"In the beginning . . .

"Then God said . . .

"By the seventh day . . .

"Now the serpent was more crafty . . ."

Whether or not a passage in the Bible is to be taken literally, it must be taken seriously. [6] This means reading it with attention to find out what it actually says. "Sit down before fact like a little child," said T H Huxley, Darwin's famous friend, "and be prepared to give up every preconceived notion." He was referring to the world of nature, but the same is true of Scripture. Those who read the Bible to pull it apart, to force their own ideas on to it or to gain support for some eccentric private opinion, are not taking it seriously any more than those who ridicule it.

This relates to the distinction between "the letter" and "the spirit" of the Law. Jesus taught two things about the Old Testament Law which, like many of his teachings, appear to be contradictory but are not. He upheld the letter of the Law ("I tell you the truth, until heaven and earth disappear, not the smallest letter, not the least stroke of a pen, will by any means disappear from the Law until everything is accomplished.") but also gave teaching starting "You have heard it was said, 'Do not do so-and-so . ." followed by "But I tell you . . ." in which he emphasised the underlying meaning (the *spirit*) of the Law. This can be condensed to two exhortations, both of which can be applied to all Scripture:

i. The *letter* mustn't be tampered with (i.e. don't be liberal)
ii. It is the *spirit* of the words that matters (i.e. don't be literal)

The Bible is universal; it is for all people, of all ages. The first chapters of Genesis have both grandeur and simplicity and can be understood, at different levels, by children, by the uneducated and by scholars. Interpretative pictures of the Genesis story can be found in Sunday school books, on the ceiling of the Sistine Chapel, and in many places in between. There is something wrong with insisting on a particular interpretation of Scripture. It is as though we are cutting Scripture down to size; our size.

I mentioned Sunday schools: many adults appear to be stuck imaginatively in the watercolour world of the Sunday school pictures, perhaps because Sunday school was the last contact they had with Christian teaching. Many atheist attacks on Christianity are written by authors who have discovered little theology since Sunday school, and who persist in attacking straw-man beliefs not held by mature Christians.

A true Nursery Tale

Here is an idea that might help to make sense of the apparent discrepancy between the first chapters of Genesis and the discoveries of science. Imagine a mother with her little boy on her lap. He is bright and asks her many questions about where he came from, and what Daddy does all day. So his mother dutifully answers him, in simple words that he can understand. These conversations happen to be recorded and, years later, the written notes are discovered. What would be the point of arguing about whether they were historical or scientific? They are just narrative. What the mother tells the child is *true* (would she lie to her little boy?) but her words are shaped for his understanding and imagination. They are not yet the whole truth because the child cannot cope with that.

Suppose that when he asked where he came from she had replied "out of my tummy." What would we make of a group of commentators who insisted that, because the words must be taken literally, a miracle must have taken place? Or another group who described the story as a myth because modern science has shown, beyond doubt, that babies come from the womb, not from the stomach?

I offer this analogy to those who realise that Genesis is meaningful but who cannot take the words literally. The first human beings must have been rather like the little boy in the story; intelligent but lacking in knowledge. [7] Think of our commonplace school subjects such as history, science, maths and geography: they wouldn't know any. If you object to the story because it might trivialise the Genesis account, don't forget that the intellectual gap between a mother and her child is far smaller than that between God and mankind.

The point of the story is not just the boy's childishness; she is giving him a *précis*. The first chapter of Genesis is also a précis. How do you condense the creation of the universe, the solar system, the earth and all its living contents into thirty-nine sentences? You can't talk about igneous rocks and pteridophytes and isopods. You have to tell a short story in simple words. The result is not trivial; it is magnificent. Think of Haydn's *Creation*, the Sistine Chapel ceiling, *Paradise Lost* . . .

The story in Genesis is not "scientific". The descriptions of the Creation are given in homely language: that of farmers, not scientists (the word "livestock" is used several times: no biologist would classify wild animals separately from livestock); words that would make sense both to the earliest human beings and to their descendants for thousands of years. This simplicity is there throughout: only seed plants are mentioned by name, and very general categories of animal (sea creatures, birds, wild animals, livestock . .), and it is wonderfully demonstrated by the throwaway remark, "He made the stars also" halfway through the first chapter.

There has been some discussion about what the word "kind" means in the first chapter of Genesis (" the fruit tree yielding fruit after his kind . . . cattle, and creeping thing, and beast of the earth after his kind . .") with attempts to make some sort of biblical Classification. One recent book suggested that "kind" was equivalent to the category "family" in the classification, in spite of the fact that there are well over a thousand biological families, and no details are given about the grouping of the animals and plants listed in Genesis. If you accept the Creation account as *story* the problems melt away. The *Good News Bible* translates the Hebrew wording like this: " so the earth produced all kinds of plants, those that bear grain and those that bear fruit . . "Let the earth produce all kinds of animal life: domestic and wild, large and small . ." Does it not simply mean *all sorts of plant* and *every type of animal*? Why do we have to make things so complicated?

The other day I found support for my mother-and-child story from an unexpected source. John Calvin said "As nurses commonly do with infants, God is wont in a measure to 'lisp' when speaking to us." (I like the idea, but the word "lisp" is wrong; simplicity is not the same as baby-talk.)

Someone might wonder, this is just a story: is there no *true* interpretation ? How do we know what actually happened at the beginning of all things? I have no idea: it is a "Newton's Dog" question. Here are some suggestions:

We must keep an open mind. This is not a black-and-white issue; as shown by the fact that clever and thoughtful people have changed their minds both from creation to evolution and vice versa).

Science is not static but changes. Its findings are provisional. The science of 100 years ago is different from that of today. Who knows what it will be like in another 100 years? If Scripture is tied to any current scientific idea it will soon become out-of-date.

No one knows what "to create" actually means. We can only make mental pictures. Instantaneous creation involves *appearance of age* (Gossery), and it is hard to know where that stops. But need creation have been instantaneous? In the Old Testament, when God "raised up" judges or prophets they were born naturally and grew up as children and teenagers just like everybody else. God used natural processes. That may be an unfair analogy for creation, but is it out of the question that the Creator used evolution of some sort during creation? (Don't forget, evolution is not the same as Darwinism.) This brings us back to the "6 days" problem: the time element divides opinion sharply. I don't know the answer.

The Genesis passages have multiple meanings. There is no interpretation agreed by all Christians, so we must be wary of linking our particular interpretation of the creation story to other doctrines. One Christian author has described belief in the first eleven chapters of Genesis as the "acid test" of belief in God, and in salvation through Jesus; another writes that inserting eons of time into Creation undermines the Gospel. I understand where these writers are coming from, but these are statements of opinion. TEs and OE creationists strongly disagree with such extensions to the interpretation of the early chapters of Genesis.

One YEC author has even stated that if Genesis is untrue, we might as well assume that no God exists at all. This sort of remark delights atheists. The writer has decided that "true" and "literally true" mean the same thing. The danger of this teaching is that young people might simply go along with it and give up their belief in God because they have changed their minds about the details of the Creation account. Incidentally, this statement is a strange one for a Christian leader to make. Is he really suggesting that a person's experience of God depends on the interpretation of a handful of verses in an ancient manuscript? He is speaking for no Christian that I have ever met.

Some Christians argue that we ought to accept the literal truth of Genesis because the New Testament apostles did; in particular because Jesus did. They quote New Testament references to Adam and Eve, and to the Flood. It is an argument which needs to be carefully thought through, instance by instance. However, because a great teacher referred to an event or person in Genesis, or used them as an example in a letter or sermon, it does not prove that he thought they were *historical* in our modern sense. Surely we just don't know. (A speaker might refer to "the patience of Job" without our knowing whether he thinks Job is historical, like Abraham, or part of an Old Testament parable.) It is wise to treat Genesis as historically true for "teaching, rebuking, correcting and training in righteousness" [8] but there is nothing in Genesis about the age of the earth, or animal death, or dinosaurs.

Scripture is only meaningful to those who take it seriously. However, the question about which parts should be taken *literally* is open, as is the question about what the word "literally" can actually mean when what is being described is beyond anyone's experience or imagination.

Notes to Chapter 14

1. *Reflections on the Psalms* (1958) by C S Lewis (Ch. XI)

2. In the book *In Six Days* at least six of the 50 featured scientists describe how they became YECs after studying evolution in depth.

3. A note just for Christians: The Bible is the *word* of God, but Christ is the *Word* of God. The former is a useful metaphor, but the latter expresses something more important. Christ is the *logos*, the expression of God Himself, referred to in the first verse of John's gospel: "In the beginning was the Word . . " Misunderstanding the importance of that capital letter can lead to *bibliolatry* (Bible worship).

4. Exodus 1. I owe this observation to Philip Yancey: *The Bible Jesus Read* (1999); Ch 1

5. There are parallels between the Old Testament stories and Christ's teaching in the New Testament which, though deep, was often expressed in the form of the stories known as parables: anecdotes from everyday life ("the prodigal son", "the good Samaritan") which made a particular point. He often used figures of speech, calling his disciples salt and light, or comparing them to vine branches. He likened himself to a good shepherd or to bread. He more than once rebuked his listeners for taking him too literally.

6. This is also reflected in the gospels. Jesus always responded to people who took him seriously, but had little or nothing to say to those who were just curious. In the last hours before his crucifixion, Jesus stayed silent in front of King Herod, who just wanted to be entertained by his famous prisoner; but he talked to Pilate, who took him seriously.

7. See C S Lewis' description of "paradisal man" in Chapter 5 of *The Problem of Pain*

8. 2 Timothy 3:16 This verse, ("All Scripture is God-breathed . . ") is familiar to Christians. It emphasises the uniqueness, sublimity and importance of Scripture.

Chapter 15

Comments on the Great Debate

Evolution is a huge subject. These next chapters look at some stray points.

Consilience

Biologists accept organic evolution, not because there are five or six telling pieces of evidence, but because it makes sense of all their studies. It is the background pattern behind biological thinking. In textbooks and pop science books, however, the same arguments and examples keep cropping up. If you didn't know better, you might think that evolution was just about explaining giraffe's necks and peppered moths; sickle-cell anaemia and some tropical finches. But it's far more than that. Behind any detailed work on genetics or geology; microbiology or plant anatomy, there is the assumption of evolution. Each new discovery or advance slots into place, in particular new work on DNA sequencing. This is "consilience" - evidences from everywhere, all pointing in the same direction without major anomalies. Here is a good description of consilience - from an unexpected source (Pope Pius XII writing about evolution half a century ago):

"It is indeed remarkable that this theory has been progressively accepted by researchers, following a series of discoveries in various fields of knowledge. The convergence, neither sought nor fabricated, of the results of work that was conducted independently is in itself a significant argument in favour of the theory." [1]

A friend of mine in America wrote this to me in a letter about YE creationism:

"By holding to the recent-creationist position, they challenge not only orthodox biology, but also *geology, geography, astronomy, cosmology and physics*. The onus

is on the recent-creationists to construct alternative versions of *all these sciences* that are more successful at making predictions than the orthodox versions of these sciences. This, I submit, they have totally failed to do."
(italics added, to emphasise consilience)

It is because of consilience that if someone claims that some one thing (an animal, an organ, a gap in the fossil record) "disproves evolution" he is not taken seriously by most scientists. It is as though someone said that a particular equation "disproves algebra".

Consilience is strong confirmation that evolution is true, but does not *prove* it. It is still possible that there is a better background explanation of the occurrence and variety of living things. No one disputes that there is a *relationship* between all living organisms but must it be an evolutionary relationship? There are other relationships. The relationship between the elements in the periodic table, or between the Alps and the Himalayas, is not an evolutionary one. Gossery, or something similar, might be true, although there would be no way of proving it. One day a new scientific explanation of living diversity might be devised which is to evolution what relativity is to Newtonian physics.

Consilience applies to other things besides evolution. Most scientists accept the billions-of-years age of the universe because data from several *different* sources agree. Most of us probably think that "Civilisation is a good thing" because there is a multiplicity of evidence. Consilience is also the overarching principle behind religious belief. A Christian doesn't believe in God because of one or two pieces of evidence, but because everything points in the same direction. C S Lewis condensed the Christian experience into one sentence, a sort of consilience-in-reverse:

"I believe in Christianity as I believe that the sun has risen - not only because I see it, but because by it I see everything else." [2]

It is because belief in God, like acceptance of biological evolution, is a matter of consilience - evidence from many different directions - that Christians are not bothered by the "knock-down arguments" of atheists. You will not persuade a biologist that evolution is wrong by arguing about the details of a feather or the sudden appearance of flowering plants in the fossil record. And you will not persuade a Christian that his beliefs are wrong by talking about earthquake victims or the extreme matching between human and chimpanzee DNA. You cannot overturn a lifetime of experience by a few specially chosen counter-examples. I say "experience" because these things are much more than just intellectual; they are convictions, part of the whole personality.

Consilience is not the same as consensus. It applies to evidence, not to interpretation; to raw rather than processed data.

What is it about Darwin? (2)

In the Bibliography there are books called *The Darwin Wars*, *Darwin on Trial* and *Darwin's Black Box*. Also *Doubts about Darwin*, *Alas Poor Darwin*, *Ever Since Darwin*, *Finding Darwin's God*, *Rescuing Darwin* and *What Darwin Got Wrong*. If you consult other books on evolution you will find countless references to Darwin or Darwinism. It is remarkable how one man has become so identified with a branch of science. Other scientists who come close are Newton and Einstein but you don't get books called *The Newton Wars* or *Alas Poor Einstein* filling the popular science bookshelves. I checked a search-engine on the Internet: there were 3000 references to Einstein, 560,000 to Newton and an astonishing 2,980,000 to Darwin, the amateur naturalist.

This is extraordinary. Why should one man become so identified with an idea in biology that a century and a half later his name is everywhere? Darwin did not originate the idea of evolution (the Greeks got there first in about 550

BC) and, under different names, it was commonplace in the nineteenth century. Natural selection was not just his idea; it was shared by several others. Lamarck knew about it, many decades before Darwin. Although he did not develop the idea into a system he recognised the power of what would later be called natural selection. He stated that because of the prodigious fecundity of the lower animals, if Nature did not step in to stop the colossal increase in numbers, the earth would become uninhabitable. And therefore the strongest would survive. [3]

In the early 1830s Charles Lyell referred to the process in his *Principles of Geology*. Edward Blyth mentioned it in papers written in the mid-1830s. Patrick Matthew was the naturalist who actually coined the term "natural selection" back in 1831. Alfred Russell Wallace produced papers, in 1855 and 1858, which showed that he had reached the same conclusions as Darwin. Wallace sent his 1858 paper to Darwin which came as a great shock to the latter, who had continually put off publishing anything. The Wallace paper and a hastily prepared one by Darwin were both read at the same meeting of the Linnaean Society in the summer of 1858, and the following year Darwin published *The Origin of Species*. Professor Hoyle suggested that Wallace's paper came to Darwin as a flash of light, "illuminating with precision ideas he had struggled with himself for almost twenty years." [4] This implied that Wallace was the real genius and that Darwin arrived at the same conclusion only after he had read Wallace's paper; or, putting it bluntly, that Darwin had pinched Wallace's idea. Darwin's biographers, Desmond and Moore, tell a different story: Darwin's ideas were fully formed as a result of many years of patient work and the fact that Wallace had come to the same conclusion was a coincidence. What Wallace's paper did was to *galvanize* Darwin into going public. [5]

This shows that the idea of natural selection was in the intellectual air throughout the nineteenth century. Malthus' *Essay on the Principle of Population* (about humanity outstripping its food supply, and the weak and improvident losing out in the struggle for existence) had been widely read. The Industrial Revolution was in full swing: the Victorians saw old machines being supplanted by better ones and progress seemed to be inevitable (they missed the point that evolution is about *change*, not *progress*). Darwin was not a lonely genius, and in many respects was not a "Darwinist".

If Darwin had become a clergyman (the "profession" originally intended for him) instead of a naturalist, someone else would have fitted into the Darwin "slot". So why did Darwin get all the credit and not the others? It was partly because he was the man on the spot (Wallace was collecting specimens on the other side of the world when their joint papers were read), partly because of the copious evidence he had collected, partly because of the support of his influential scientific friends like T H Huxley, but mostly for the reason his son Francis gave many years later:

"In science the credit goes to the man who convinces the world, not to the man to whom the idea first occurs." [6]

Fair enough; that makes good sense. But those events were a century and a half ago. Why is his name still everywhere today? (Recently I read a newspaper article about "Evolution and Faith" and Darwin's name cropped up 53 times! A modern genetics article would not have 53 references to Mendel.) It has not much to do with biology: many of Darwin's ideas have been superseded. Instead it reflects the philosophy of Darwinism; a superstructure piled up on the more modest foundations of his work. A recent article in *New Scientist* referred to *The Origin of Species* as "perhaps the most important book ever written." The word "perhaps" redeems the opinion from sheer absurdity, but the remark itself demonstrates that Darwinism is not just a theory in biology but a philosophy.

Many would go further, and refer to it as a *religion*. Evolutionists fiercely deny this but may be referred back to a comment by Julian Huxley (part quoted in Chapter 5) in the speech he made at the Darwin centennial celebration in 1959:

"In the evolutionary pattern of thought there is no longer either need or room for the supernatural. The earth was not created, it evolved. So did all the animals and plants that inhabit it, including our human selves, mind and soul as well as brain and body. So did religion . . .

"Finally, the evolutionary vision is enabling us to discern, however incompletely, the lineaments of the new religion that we can be sure will arise to serve the needs of the coming era." [7]

When God is dethroned something is always put in his place, and evolution (or "Nature" or even the gene) is often spoken of with awe. And as Mohammed is to Islam, so Darwin is to Atheism - no one must speak a word against the prophet. That is why when any scientist supports creationism, or questions the creative power of natural selection, the scientific establishment issues a *fatwa* against him or her. There are instances of scientists losing their jobs, or failing to secure them, because they allowed their doubts about evolution to become known.

Darwin's name is often linked with those of Marx and Freud; a trio of Victorian thinkers whose views changed the world. James Le Fanu has pointed out [8] that they had several things in common: each had an important insight (or Big Idea); each denied the reality of the self as an autonomous being; each was openly, or secretly, hostile to religion. Their schemes became overarching theories intended to explain *everything*. But they cannot: they are partial explanations, not universal. So Marxism is in decline and Freudianism is flawed . . how soon will it be before Darwinism follows them off the stage ?

Meanwhile, as the book title puts it, "Alas Poor Darwin!" His name is dragged to and fro; often through the mud; a name hated and feared by many religious people and hijacked by others as a banner for nasty '-isms'. Creationists refer to him with contempt. Some atheists almost worship him: I have a book with prominent photographs of his chair and table, and in a recent bicentenary programme a well-known evolutionist showed us the very piano keys that Those Fingers touched. These polarised views are absurd. Darwin was a late Victorian gentleman; shy, anxious and continuously ill; an outstanding naturalist and hardworking author with important ideas, a clear mind, and a splendid streak of poetry. Why should he have all this done to him? Alas, poor Darwin.

Tit for Tat arguments

J B S Haldane once remarked : "Even the Archbishop of Canterbury is 65 percent water." Haldane was much given to attacking "religion", and probably thought his teasing remark was very witty. In a sense, it is. Many of us find vestments and regalia slightly comic, and a Marxist like Haldane would have hated robes and rituals. But this particular remark was pointless, because Haldane himself was 65% water too and so is everyone else, including both clergymen wearing vestments and scientists wearing doctoral gowns. Personal taunts and *ad hominem* remarks are useless in argument. (An *ad hominem* remark attacks someone's character instead of answering that person's argument.) We may suspect that the anger and rudeness of X has a psychological explanation, but that has nothing to do with the logic of his or her arguments. It is *ad hominem* if anyone attempts to give reasons *why* X holds those views, instead of discussing the views themselves. It is a change of subject, often used by people when they are losing the argument. Instead of discussing the facts, people will say things like, "You only say that because you are a socialist!" or "That's a typical woman's remark!"

Huxley's statement is also a good example of reductionism in its most pointless form. Next time you hear someone saying that men and women are only naked apes, or that thought is just brain chemistry, remember that the speaker is 65% water.

Many arguments about evolution or religion can, like Huxley's jibe, be turned against the persons making them. Religious people are said to need the "consolations of religion" (the comfort that comes from belief in God): therefore, we are told, their belief is wish-fulfilment. A retort is that the boot is on the other foot; that it is the *denial* of God, by those who can't bear the thought of Him, which is wish-fulfilment. Atheists liken religious belief to a virus that infects human minds; their religious opponents simply reply that it is atheism which is the virus. Is there a "meme" (an idea which multiplies almost by natural selection) for religion? Well, so is there a meme for atheism. Is there a special region of the brain where atheism is located? Guess what! it is exactly the same region where theism is located (someone called it the "God module"). And so on: all these arguments are hopelessly double-edged; they simply cancel each other out.

A recent double-edged argument has been called the "argument from personal incredulity", a phrase from one of Richard Dawkins' books. [9] If a creationist says "I cannot believe that something as complex as the human eye could have evolved by natural selection" the retort is that matters of fact do not depend on what a person happens to be able to believe. But this put-down is useless because it applies to *any* confession of disbelief, including that of the evolutionist who calls special creation "clearly incredible", the materialist who "cannot believe" in the supernatural, or the geologist who disbelieves that most sedimentary rocks were laid down during Noah's flood. [10] Note: in the sport of creationist-bashing, YECs are attacked on two incompatible fronts. At one moment they are ridiculed for their personal incredulity and the next they are accused of "believing six impossible things before breakfast." They don't seem to mind very much.

Creationism is *false*, we are told, because the evidence is against it. But it is also *unfalsifiable* because it depends on miracle. Evolutionism is *false*, we are told by others, because it is an extension of science which has no empirical justification. But it too, is *unfalsifiable* because natural selection is so formulated that it can explain anything at all.

Professor A's thoughts are predetermined chemical events in the neurons of his brain, so why should I attend to them? Doctor B's ideas are conditioned reflexes resulting from his strict upbringing, and therefore unsound. The Reverend C has been brainwashed. Anyway, all three of them, A, B and C, consist of 65% water.

D is an atheist because he was brought up as an atheist. E is an atheist because he was brought up in a Christian home, then rebelled against his parents. Miss F was educated in a convent: of course she is religious - what chance has she got?

"You say that because you are a creationist!"

"You only say that because you are an evolutionist!"

This is very tiresome and silly; unfortunately, it is one of the reasons why the Great Debate is still so polarised. Keep an eye out for these self-cancelling arguments.

The words "evolution" and "natural selection"

Evolution is a loaded word. It has several different meanings and is heavy with associations. Darwin only used it a few times in *The Origin of Species* towards the end. The verb "evolved" is the last word of the book:

"There is grandeur in this view of life, with its several powers, having been originally breathed by the Creator into a few forms or into one; and that, while this planet has gone cycling on according to the fixed law of gravity, from so simple a beginning endless forms most beautiful and most wonderful have been, and are being, evolved." [11]

Evolution is a slippery term meaning different things to different people. Phillip Johnson is a lecturer in law, and really does know how words are used and misused in argument. He writes:

"Much confusion results from the fact that a single term - "evolution" - is used to designate processes that may have little or nothing in common. Shift in the relative numbers of dark and light moths in a population is called evolution, and so is the creative process that produced the cell, the multicellular organism, the eye and the human mind. The semantic implication is that evolution is fundamentally a single process, and Darwinists enthusiastically exploit that implication as a substitute for scientific evidence. . . . The vocabulary of Darwinism inherently limits our comprehension of the difficulties by misleadingly covering them with the blanket term evolution." [12]

This is from an early chapter in Johnson's book *Darwin on Trial*. He makes the same point in the epilogue:

"Evolution" stands for the modest knowledge that science actually has gained about how organisms vary, and also for the vast naturalistic creation story about how mutation and selection brought life to its present complexity. Do you admit or deny the "fact of evolution"? Deny it and you seem to be denying that island species vary from mainland ancestors, or that dog breeders have produced St. Bernards and Dachshunds from an ancestral breed. Admit it and you are taken to have admitted, quite without support in the evidence, that an ancestral bacterium changed by a vast series of purposeless adaptive steps to produce today's whales, humans, insects, and flowers. If "evolution" is assumed to be a single process, then to admit any aspect is to admit the entire story." [13]

Another reason for being wary of the word "evolution" is that it is often used to give a spurious scientific air to a statement. "The robin has evolved a red breast" tells us no more than "the robin has a red breast" but it sounds grander. Popular natural history programmes often have the word "evolution" scattered about in this way, as if to establish the credentials of the presenter.

A third reason is that "Evolution" is often personalised. As I reported in Chapter 2, one of our videos had the line ". . . evolution has *seen to it* that Amy is born with a few essential abilities . . . to give her a good start in life." An author writes of natural selection "wasting its time" and another has asked what would happen if natural selection "does not like what's going on." An article in *New Scientist* referred to "the extent to which evolution likes to simplify matters." These sentences are from just one book:

"evolution recruits what it needs .."

"natural selection cleverly assembles workable structures .."

"natural selection polices its regime with diligence and brutality .."

(My original list had thirteen examples; I spare you the others.) Some readers would find this sort of writing entertaining but it is unscientific. Worse, there is a danger that unsophisticated readers will start to believe that there is a real Mother Nature, with a mind of Her Own, who is running everything.

OK, these are just pieces of scientific journalism, but they promote the false idea that evolution and selection can actually *do* something, that they are *agents*.

The term "natural selection" is unfortunately anthropomorphic. To select is to choose, and only a mind can choose. The imaginative gap between the two ideas "nature selects" and "Mother Nature selects" is tiny. It is not only creationists who oppose this way of looking at evolution. In their recent book *What Darwin Got Wrong* [14] Jerry Fodor and Massimo Piatelli-Palmarini (who are secular humanists) strongly oppose any suggestion that evolution is an *intentional* process and insist that Darwin made a mistake in likening natural selection to humans doing selective breeding. They are as fiercely against Mother Nature or Selfish Genes as they are against God.

Evolution is for some a substitute religion, and it is often over-written. We have already seen what has happened to poor Darwin. An article about wildlife is called *Wonders of Evolution* but there are no articles on the solar system called *Wonders of Gravity*, and when we praise someone's baby girl we don't call her a Wonder of

Embryology. "Evolution" is just an abstract noun, like "architecture" or "mechanics".

Evolution is a particularly unpleasant word to creationists because they associate it with struggle, cruelty and waste. This is unfortunate. We saw in Chapter 10 that evolution is no more *cruel* than nature itself. (Not "Herself", note.) As for *waste* I suggest that the use of such a word demonstrates a failure of imagination. Consider these statements:

This baby fox has just been run over and killed: what a waste!

For every successful organism there are countless failures.

This tortoise lived to a ripe old age.

The dinosaur design was ultimately no good; they all died out.

People who make this sort of statement are looking at the world through a sentimental, anthropomorphic lens. An animal that lives for 20 years is not twice as valuable as one that lives for 10 years. A plant with flowers is not more "successful" than a fern with spores. Trilobites and titanotheres were not *failures* because they died out. Nature is itself: it is not referable to, nor understood by, us. Most of nature has never been seen by human eyes, and never will be. Does that mean that it's wasted? There are no scales on which we can weigh these matters.

The main reason that the word "evolution" is disliked by creationists is that it is associated with Godlessness, and disregard for the Bible. When a theistic evolutionist tells a creationist that God "used evolution" there is always a negative response. This is unfortunate: evolution is not the same as neo-Darwinism. The latter is a proposed *mechanism* for the former; one which may be reaching the end of its shelf-life. We might narrow the gap between evolutionists and creationists by calling evolution "development" - a neutral word that leaves open the question of a mechanism.

Neo-Darwinism is a paradigm; the Establishment View, [15] fiercely defended against criticism. Its terminology, of mostly abstract nouns, has channelled ideas into grooves which prevent lateral thinking. But what if the grooved ideas are wrong?

As for "natural selection", it is a phrase often used as a stop-gap. Surely, most animals and plants die, not because they are ill-adapted, but because of the ordinary chance events of life: an animal in the wrong place at the wrong time, or a seed blown out to sea. Selection is too easily invoked: we glibly say that if an organism lives, it is "selected for"; if it dies, "selected against." If a population changes, that is directional selection; if it doesn't change, it is stabilising selection. The snail has a shell: that is the product of natural selection; a slug has no shell: that, too, is the result of selection. Is there anything that cannot be explained by natural selection? The idea of natural selection is not *wrong*; but half the time it is superfluous and most of the time it is guesswork. Evolution happened in the past and *no one will ever know* what caused it.

Here are two questions to think about:

Is the phrase "God used evolution to create living things" equivalent to "My mother used nutrition to feed her family" or "Renoir used impressionism to make his paintings"?

Why is it that someone will say "The heron evolved a long beak" but would never say "The bulldog evolved a snub nose"? Natural selection and artificial selection are supposed to be equivalent processes.

Notes on Chapter 15

1. From *Humani Generis* (1950) cited by Stephen Jay Gould in *Rocks of Ages*, Chapter 2

2. *They asked for a Paper* (1962) "Is Theology Poetry" by C S Lewis

3. From *The Evolution of Living Things*, Chapter IV, "Lamarckism" by Graham Cannon, 1958. Textbooks contrast Lamarck (hopelessly wrong) and Darwin (brilliantly right), but Lamarck knew about natural selection, and Darwin accepted a form of Lamarckism ("use inheritance") throughout his writing!

4. *The Intelligent Universe* (1983) by Fred Hoyle, Chapter 2,

5. *Darwin* (1991) by Adrian Desmond and James Moore

6. *Eugenics Review* (April 1914) Sir Francis Darwin

7. *Evolution after Darwin*, (1960); the University of Chicago's record of the Centennial Celebration (Vol. 3)

8. *Why Us ?*(2009) by James Le Fanu, Chapter 10,

9. In his book *River Out of Eden: A Darwinian View of Life* (1995), Richard Dawkins told us that we should never take seriously anybody who says that he cannot believe that something or other could have evolved by gradual selection. He called such a view "The Argument from Personal Incredulity" and remarked that it was frequently the prelude to "an intellectual banana-skin experience".

10. The book *But That I Can't Believe!* (1967) by John Robinson, Bishop of Woolwich was based entirely on "Personal Incredulity." He went through Christian core beliefs, explaining why he couldn't accept them. (He stayed on as a bishop, though.)

11. *The Origin of Species*, Chapter XV.

12. *Darwin on Trial* (1993) by Phillip Johnson, Chapter 5

13. *Darwin on Trial*, Epilogue

14. *What Darwin Got Wrong* (2010) - see Bibliography

15. At a creationist meeting I attended, the speaker was obliged to announce that his university dissociated itself from his views! Surely that's very strange. We don't blame Oxford if one of its dons is an atheist, so why should we blame Leeds if one of its professors is a creationist?

Chapter 16

Comments on the Great Debate (2)

Science and religion

["The time has come to break the silence and restore a coherent, balanced view of ourselves and our world by putting aside biology's foundational evolutionary theory and embracing the dual nature of reality"
James Le Fanu; *Why Us?*

"I found it difficult to imagine that there could be a real conflict between scientific truth and spiritual truth. Truth is Truth. Truth cannot disprove truth."
Francis Collins; *The Language of God*]

Ultra-Darwinists try to widen the "evolution-versus-creation" debate into a "science-versus-religion" debate. In a recent television programme Intelligent Design was referred to as an "attack on science" even though nearly all the IDers in the programme were working scientists. Then there is that claim in a recent book (see Chapter 10) that ID "imperils American global dominance in science" - an odd claim when you remember that ID is about bacterial flagella or the evolution of a bird's lung.

Professor Lewontin has pointed out that science is not just a collection of facts about the world, but is "the body of assertions and theories about the world made by people who are called scientists." [1] Science is a social activity and will reflect the opinions and prejudices of those taking part. This is particularly true of a historical science like evolution where so much is guesswork.

It is sometimes stated that a scientist, after long training in objectivity, cannot hold any religious beliefs honestly. I have been protected from this mistaken view by sheer weight of experience. Nearly all my school science teachers, and several of my university lecturers were Christians. During my middle years as a teacher at Millfield the Head of Science was both an outstanding chemist and a committed member of the Plymouth Brethren; he was often to be found preaching in the open air down in the town. Two different Heads of the physics department were deacons in the local Baptist Church. The three elders in our church in Glastonbury were trained as scientists; a doctor in general practice, an ophthalmologist and a bacteriologist. The pastor of a big Church near Wells is a nuclear physicist and worship is led there by a chemist. A month or two ago I went to a lecture in the Bishop's Palace in Wells by an astrophysicist who is also an ordained minister. These people are enthusiastic Christians and enthusiastic scientists. To quote again from Dr. Collins' book: "The God of the Bible is also the God of the genome. He can be worshipped in the cathedral or in the laboratory. His creation is majestic, awesome, intricate, and beautiful - and it cannot be at war with itself."

There is no conflict between "science" and "religion" and never has been. The early scientists were all religious men, and many leading scientists are religious men and women today. The famous set-piece *Science-Versus-Religion* tussles from history were not what they seem. The story of Galileo and the Catholic church is nearly always presented as a struggle between a heroic scientist and an obscurantist Church. But the whole business was far more complicated than people realise. Stephen Jay Gould tells the story in *Rocks of Ages*. [2] He describes Galileo as a "brilliant hothead who had caused trouble before" and who had gone out of his way to annoy the Establishment until they finally lost patience with him. It was not primarily a science-versus-religion dispute at all.

Another famous confrontation was that between T H Huxley (representing Darwin) and Bishop Wilberforce (representing "the Church") which took place in Oxford in 1860. Like the Galileo episode this has passed into folklore and the details in the books are mostly wrong. At

Millfield we had a video, *Darwin's Bulldog*, which gave a dramatised version of the debate, with Wilberforce played as an ignorant cleric and Huxley as a brilliant scientist. The controversy was presented as a fascinating story, with the protagonists scoring debating points off each other in the course of a theatrical exchange of arguments. You would never guess from the film that most of it was pure invention. No notes of the Oxford meeting had been kept. Both sides thought that they had won the argument. The film made out that it was a black-and-white issue, with the religious side losing. But there was no "religious side." Wilberforce may have been a bishop, but he was also the vice-president of the British Association for the Advancement of Science, and he used arguments supplied to him by other scientists. In *God and the Biologist* Professor Berry writes:

"It is unfortunate that Wilberforce was a Christian and Huxley was not, because the evolution debate became a religion *versus* science argument from that moment on." [3]

The famous Scopes Trial, in Dayton, Tennessee back in the 1920s was only partly a science-versus-religion dispute. The leading prosecutor, William Jennings Bryan, was not a Genesis literalist. According to Phillip Johnson, "He opposed Darwinism largely because he thought that its acceptance had encouraged the ethic of ruthless competition that underlay such evils as German militarism and robber baron Capitalism." [4] The 'scientists', the Scopes defence team, made a poor case for science. They relied heavily on dubious pre-human fossil evidence including that for the so-called "Nebraska Man" (a misidentified pig tooth) and the famous "Piltdown Man" (later discovered to be a fossil fraud).

There is no necessary conflict between science and religion in general, and TEs also claim that there is no reason for conflict between organic evolution and the creation accounts found in the Bible. It depends on your definition of the word "evolution" and your interpretation of the Genesis passages. The real conflict is not between two *scientific* positions but between two worldviews, evolutionism and creationism, which are moving further apart: a separation with which the ultra-Darwinists and the YE creationists are very content. Recently I heard a speaker say of the Creation, "There are only two views: EITHER God did it OR Evolution did it!" (A possible cartoon came into my mind: two sparrows on the Eiffel Tower in Paris discussing how it got there: "There are only two views: either Eiffel did it or Engineering did it!")

Science is a by-product of the so-called "Enlightenment"; the way of thinking in which reason and logic are held to reign supreme over other ways of using the mind. That is fine for hard science, but there are many aspects of being human in which simply applying logic or the scientific method are actually *un*reasonable: in art and other products of the imagination, in personal relationships like love, and in religious experience. These are not illogical but supra-logical. A maniac, Chesterton told us, is mad not because he has lost his reason but because he's lost everything except his reason. [5] The Enlightenment did not *introduce* reason into the world, but isolated and inflated it. Where there is a conflict between science and religion it is often because scientists try to apply scientific tools to non-scientific problems; to "render unto Caesar the things that are God's." Scientists, like everyone else, are human beings, not robots; they think as they do partly because of what they have *experienced*. If it wasn't for your experience you would not know what you are talking about, or what words mean. Now, when a materialist talks about religious belief he is coming to it "from below"; he has no experience of what he is describing. He is like a colour-blind man who criticises colour vision.

Mind and matter

". . . a theory that is the product of a mind can never adequately explain the mind that produced the theory."
Phillip Johnson, *Reason in the Balance*

The Genesis account is the Creation Story of the Jewish religion, accepted by most Christians. The specifically Christian story is much shorter, and found in the first few verses of the Gospel of John:

"In the beginning was the Word, and the Word was with God, and the Word was God. He was with God in the beginning. Through him all things were made; without him nothing was made that has been made. In him was life, and that life was the light of men. The light shines in the darkness, but the darkness has not understood it"

This famous passage contains both "vertical history" and heightened language, so cannot be directly compared with any scientific account of origins. There are no details of what happened at Creation. But it answers a fundamental question which lies behind all the creation stories: "Which came first, mind or matter?" By *mind* I mean as humans understand it: self-consciousness, reason and imagination; not just brain activity.

Materialists believe that mind arose from matter during the course of evolution. (Although one or two keep the door open to the idea of a Cosmic Mind which is somehow part of matter and was there 'all along' - whatever that means. [6]) I don't know what philosophers think; but I doubt if it is a matter of biology. It never cropped up on any syllabus, although animal *behaviour* was included in both our university courses and our classroom teaching. There are two reasons why mind cannot arise from matter. Firstly, irrationality cannot somehow *develop into* rationality; that is not how reason works. Secondly, chemical reactions and logical arguments are in totally different categories: one cannot turn into, or produce, the other. It is surprising that materialists claim consciousness as part of their system, because it is so clearly *immaterial*.

J B S Haldane stated in his essay *Possible Worlds*:

"If my mental processes are determined wholly by the motions of atoms in my brain, I have no reason to suppose that my beliefs are true . . . and hence I have no reason for supposing my brain to be composed of atoms." [7]

In a later essay, he wrote:

"I am not myself a materialist because, if Materialism is true, it seems to me that we cannot know that it is true. If my opinions are the result of the chemical processes going on in my brain, they are determined by the laws of chemistry, not those of logic." [8]

What could be more sensible? (I can't quite make Haldane out: he was a Marxist but claimed that he was not a materialist. Perhaps dialectical materialism isn't materialism after all. Perhaps he was just teasing.) Haldane's statements make excellent sense. A philosopher who claims that reason "evolved" from an irrational precursor is cutting off the branch on which he sits, for he is denying the validity of his own thoughts. This is what Haldane has put so well.

The question of reason evolving from unreason was raised many times by C S Lewis in his varied writings. This passage echoes Haldane's remark:

"The theory that thought is merely a movement in the brain is, in my opinion, nonsense; for if so, that theory itself would be merely a movement, an event among atoms, which may have speed and direction but of which it would be meaningless to use the words 'true' or 'false'. [9]

Lewis described evolutionism as a "Great Myth". He wrote:

"To reach the positions held by the real scientists . . you must treat reason as an absolute. But at the same time the Myth [popular Evolutionism] asks me to believe that reason is simply the unforeseen and unintended by-product of a mindless process at one stage of its endless and aimless becoming. The content of the Myth thus knocks from under me the only ground on which I could possibly believe the Myth to be true. If my own mind is a product of the irrational - if what seem my clearest reasonings are only the way in which a creature conditioned as I am is bound to feel - how shall I trust my mind when it tells me about evolution?" [10]

In an entertaining passage, Philip Toynbee deals with the same problem: to assert that a ball of flaming gas has produced (or evolved to make possible) the philosophy of Plato is as nonsensical as "to say that a windmill has changed into yesterday afternoon !" He goes on to state:

"Normally rigorous minds have been deceived into this gross confusion of incompatible linguistic categories by the sheer *length of time* which this change of matter into mind is supposed to have entailed. But whether the change is 'quick' or 'slow' according to our own arbitrary standards of time, it is clear that the linguistic absurdity remains." [11]

Were Lewis and Toynbee biased because they were not scientists? Here are the views of a physicist, John Polkinghorne:

"If we are caught in the reductionist trap we have no means of judging intellectual truth. The very assertions of the reductionist himself are nothing but blips in the neural network of his brain." [12]

The mathematician John Lennox quoted these words and added:

"There is a patent self-contradiction running through all attempts, however sophisticated they may appear, to derive rationality from irrationality. When stripped down to their bare bones, they all seem uncannily like futile attempts to lift oneself by one's bootstraps, or to construct a perpetual motion machine." [13]

The distinction between mind and matter is not the same as the question of the evolution of the brain. Theistic evolutionists accept that brains have evolved. They do not share the view of Wallace who believed, much to Darwin's distress, that the human body had evolved from an apelike body, but that the human brain had not. Materialists believe that the mind is a property of the brain; that when you have fully described the physical workings of the brain there is nothing called "mind" left over. Christians, and other theists, believe that Mind (or Reason, or *Logos*) preceded matter, and that our own human minds share it. You can label this view "supernatural" if you like, but there is nothing spooky about it. It is as matter-of-fact as thinking.

Here are two statements to think about:

"Scientific naturalists insist, paradoxically, that the cosmos can be understood by a rational mind only if it was not created by a rational mind."

(Phillip Johnson)

"If the human mind was simple enough to understand, we'd be too simple to understand it."

(Edith Sitwell)

Discontinuities in Creation

There are three discontinuities in the origin (or genesis) of Nature:

The Origin of Matter

There was nothing - and then matter and energy were there. Scientists and believers agree about this, whether referring to the Big Bang or the words of Scripture: "In the beginning God created the heavens and the earth." This first discontinuity has attracted Latin phrases such as "Creation *ex nihilo*" or "Creation *de novo*" which both mean that there was nothing there before the universe began.

The Origin of Life

There are two categories of object in the world, those that are alive and those that are not alive. The change from living to non-living is a one-way process. Scientists have been demonstrating this for centuries. Famous examples are Redi's experiments with maggots in 1668 and Pasteur's research with microbes in 1862, but countless experiments since have confirmed the same fact. We cannot predict what scientists will discover in the future, but so far no one has got even near to "making life" (abiogenesis).

This fact has also attracted Latin maxims, from "*omne vivum ex ovo*" (all life is from eggs) to the improved version "*omnis cellula e cellula*" (all cells arise only from pre-existing cells).

The Origin of Mind

This discontinuity is not (necessarily) between humans and animals, but between the material and the immaterial. We only know about it by subjective experience. There is a world of neurons and potassium ions in the brain, and there's a completely different world of logic; and of truth, beauty and goodness.

We cannot say anything useful about the "higher" animals here, because no one has ever experienced what it is actually to *be* a dog or a gorilla. If one day it is somehow discovered that a chimpanzee can do mental arithmetic or enjoy Mozart, the discontinuity will still be there; not between us and the chimp, but within its "mind": between the chemical activity of its brain and its immaterial thoughts.

These discontinuities are not *gaps*, as in the "God-of-the-gaps" taunt. A gap is a missing link in a chain. Instead of a chain think of a staircase with huge steps. The steps are discontinuous but there is no gap between them.

These three discontinuities can be observed and are thus part of science, the book of God's works. Three discontinuities are also recorded in the Creation Story of Scripture, the book of God's word. They do not exactly coincide with the first three (very little in the Two Books provide easy one-to-one correspondences). They are associated with the Hebrew word *bara*; this is used just three times, and is taken to mean "create" in the divine sense: i.e. something totally new has entered the world.

In the words of one Bible Commentary, "*Bara* is found in verses 1, 21 and 27; that is to say, at the beginning of all existence; at the beginning of all animate existence; and at the beginning of all spiritual existence, so far as this world is concerned." (The choice of words is interesting: "animate" and "spiritual" could be argued about. There is also a parallel with a passage from Professor Thompson's *Introduction* to Darwin's *Origin of Species*: "Between the organism that simply lives, the organism that lives and feels, and the organism that lives, feels and reasons, there are, in the opinion of respectable philosophers, abrupt transitions corresponding to an ascent in the scale of being, and they hold that the agencies of the material world cannot produce transitions of this kind. [14])

These discontinuities are a matter of fact They cannot be explained by materialism.

Christians know about a fourth discontinuity; hard to set down in clear words because "mind", "soul" and "spirit" are used with varied meanings. The fourth has nothing to do with biology and applies only to humans. It is between "Mind" in the sense of reason and "Spirit" in the sense of awareness of, and response to, God. This discontinuity is a main theme in Paul's teaching. [15]

Variations on a theme

Stephen Jay Gould argued that imperfections in nature were evidence for evolution. He wrote:

"Orchids manufacture their intricate devices from the common components of ordinary flowers, parts usually fitted for very different functions. If God had designed a beautiful machine to reflect his wisdom and power, surely he would not have used a collection of parts generally fashioned for other purposes. Orchids were not made by an ideal engineer; they are jury-rigged from a limited set of available components. Thus they must have evolved from ordinary flowers.

"Odd arrangements and funny solutions are the proof of evolution - paths that a sensible God would never tread but that a natural process, constrained by history, follows perforce." [16]

He returns to this theme later in another essay:

"We infer history from imperfections that record constraints of descent. The 'various contrivances' that orchids use to attract insects and attach pollen to them are the highly altered parts of ordinary flowers, evolved in ancestors for other purposes. . . If God wanted to make insect attractors and pollen stickers from scratch, he would certainly have built differently." [17]

These statements are a remarkable departure from Gould's normal wise argumentation. The idea that he knows better than God how to design a flower takes us deep into "Newton's dog" territory. Many lesser authors make the same mistake. One wrote that "God surely wouldn't have done" something or other; another that "a sensible designer would never have used four different DNA codes" for four similar poisons; yet another that God wasn't very good at design because so many animals and plants had become extinct! Several recent evolutionist books have majored on this question of "poor design" citing the backwardly-wired retina of the vertebrate eye, or the course of the recurrent laryngeal nerve in the neck, or even the fact that people get back-ache. These are astonishingly poor arguments. Firstly they are theological, not remotely scientific. Secondly you would have to be in a very contrary mood to pour scorn on the design of the eye, which most people regard as a miracle of engineering. Thirdly these statements imply a remarkable poverty of imagination. Those who make them must have a tiny, Sunday School, conception of God.

There is a different way of looking at apparent imperfections in nature. In the first chapter of Genesis we are told that human beings were created in the image of God. This obviously does not refer to our bodies (God is a spirit) but to our mental and spiritual attributes. One of these is our creativity. We make things because the Maker has caused us to be makers: writers, artists, composers, builders and so on. If we make something well, we are following the Master's pattern. If I have made something badly, I have drifted away from source (trying to do it "my way").

If I am right, our best human creativity is like that of God. So it is not in the least surprising that "Variations on a Theme" should be such a large part of both human and divine creativity. Some of the finest art ever made consists of variations on a theme. Think of any fugue, the *St Anthony Chorale*, the *Enigma* variations, Hokusai's *37 Views of Mount Fiji*, most pieces of Classical architecture and all good poetry. The combination of *restriction* (the theme) and *diversity* (the variations) somehow canalises and enhances the creativity.

The idea of variation on a theme is behind the world of collecting and classifying (briefly referred to in Chapter 2). The fascination of collecting is that the objects are both "the same" and different. They can be grouped into categories, whether we are collecting stamps or butterflies. Many collections and classifications are of living, or once-living, organisms. Gould described biologists who follow a number of intellectual styles, pointing out that some delight in diversity for its own sake, spending a lifetime describing intricate variations on common themes, whereas others look for the underlying unity behind the differences in order to make a tidy classification. In short, some biologists love the theme, others the variation. What is the Linnaean Classification but a gigantic instance of variations on a theme? Evolution is one particular way of making sense of it.

It is strange that, in the light of this insight, Gould missed the point. One of the reasons that the flower of the bee orchid evokes wonder and admiration is precisely *because* it consists of a new device made from the same old parts. The flower becomes much more interesting (and more beautiful) to someone who has learned a little botany.

Haldane used to joke that the one thing he knew about God was that he was "inordinately fond of beetles." (There are well over six hundred thousand known species of beetle in the world; the true number is probably in the millions.) Haldane's remark was meant to be mocking, but I have always taken it to be words of praise. Why should there be any limits to the variety proceeding from the very source of creativity? After all, *some* animal group has to have the greatest number of species; why should it not be the beetles? [18]

Notes on Chapter 16

1. *The Doctrine of DNA* (1991) by R C Lewontin (Ch. 5)

2. *Rocks of Ages* (2001) by Stephen Jay Gould (Ch. 2)

3. *God and the Biologist* (1996) by R J Berry (Ch. 3). For a detailed account see *Fabulous Science* by John Waller (Ch 10) The Galileo episode and the Oxford Union debate are also described in Chapter 2 of *God's Undertaker* (2007) by John Lennox.

4. *Darwin on Trial* (1991) by Phillip Johnson (Ch.1)

5. See *Orthodoxy* by G K Chesterton (Ch. 2) A similar idea is in *The Abolition of Man* by C S Lewis (Ch. 1)

6. For example, J B S Haldane wrote: "There is a great deal of evidence that the universe as a whole possesses certain characters in common with the human mind." (*Fact and Faith*) 1934.

7. *Possible Worlds & Other Essays* (1927)

8. *Fact and Faith* (Thinker's Library No 44); Essay 5 "Some Reflections on Materialism" (1934)

9. *Transposition and Other Addresses* C S Lewis (Ch. I) The fullest discussion of the source of reason takes up the first 6 chapters of *Miracles: a Preliminary Study* (1960)

10. "The Funeral of a Great Myth" from *Christian Reflections* (1967).

11. from *Towards the Holy Spirit: a Tract for the Times* (1973) by Philip Toynbee

12. *One World* (1986) by John Polkinghorne.

13. *God's Undertaker* (2007) by John C Lennox.

14. "*Introduction*" *to the Origin of Species* (1956) by Professor W R Thompson FRS

15. e.g. 1 Corinthians (Chs. 1 and 2) in which Paul contrasts earthly wisdom with the "things of the spirit."

16 *The Panda's Thumb* (1981) by Stephen Jay Gould

17 *Hen's Teeth and Horse's Toes* (1983) by Gould

18. Compare this with the idea that both embryology and evolution are modular (remember the Lego bricks). Modularity and variations-on-a-theme are linked ideas.

Chapter 17

"Modern biology has occasioned a most unfortunate polarization of opinion between materialist evolutionists on the one hand and fundamentalist creationists on the other. Both sides tend to conflate evolution with Darwinism. The evolutionists argue that to attack Darwin is to attack evolution itself, while the creationists argue that since Darwin is faulty, evolution is wrong."

Augros & Stanciu [1]

Easing the situation

Is there any way we can help to close the gap between Darwinists and creationists? For a start, the Two Books (of science and Scripture) give *complementary* accounts of the origin and diversity of living things. This does not mean that one is more right than the other. Think of two contrasting descriptions of a work of art. A scientific account of a great painting would describe the chemistry of the pigments and the physics of how light passes through the surface glazes that the artist has applied. These would be of little interest to a connoisseur, who would want to discuss the composition, the brushwork or the inspiration of the painter. These descriptions are not opposed to each other but just different. In the Great Debate science and Scripture are answering different questions. So "Theistic Evolution" should be the answer: TEs observe the principle of complementarity by accepting both the Bible and the findings of science. But their views are unpopular with creationists, partly because they do not fit into a literal interpretation of the first chapters of Genesis, and partly because they are seen as one-sided. Wherever there is a clash between evolution and the Bible the former, it seems, always wins. The first chapters of Genesis are interpreted in the light of Darwinian theory and the stories of Adam and Noah are interpreted in the light of secular archaeology. There is no give-and-take; or, rather, Genesis does all the giving and evolution does all the taking. In response, the TEs, in an agony of reasonableness, implore the creationists to accept that there is good evidence for evolution, and that it is not somehow un-Christian to believe that God used evolution to create. But the creationists refuse to budge an inch.

This problem of reconciling two complementary accounts seems to apply particularly to adults: pupils in the classroom mostly have no difficulty in accepting that there might be a "scientific" and a "religious" account of the same thing. Is it because young people do not feel the need to have an air-tight, logical and immediate explanation of everything they encounter? If you realise that there are lots of things you don't know, competing theories don't bother you much. But if you are absolutely sure that you know everything of importance, and that you are *right*, then opposing views are clearly wrong, and must be attacked or ridiculed.

This issue can never be fully settled but the situation may be eased. A good start is to "go back to source." In science, we should accept only hard data and empirical facts. Much of evolution theory, especially that dealing with the remote past, is guesswork; what Professor Thompson, in his *Introduction* to the Everyman Edition of *The Origin of Species*, referred to as "the fragile towers of hypotheses based on hypotheses, where fact and fiction intermingle in an inextricable confusion". [2] We must look at the actual *evidence*: the details of the fossil record, and the evidence from DNA sequencing and embryology; and not just take the word of popularisers, especially those with an axe to grind. In the same way, we should go back to the actual words of Scripture, rather than to someone's interpretation of them. Some creationist teaching, particularly that based on "Flood Geology", can be persuasive, but is not actually found in the Bible.

If evolution means *NDS evolution* and creation means *YEC creation* then these two cannot be reconciled. Let us look at simpler ideas of evolution and creation.

Non-Darwinian evolution

There is plenty of evidence for biological evolution (listed in Chapter 2) in the sense of development or, in Darwin's original phrase, "descent with modification." On the other hand, there is little evidence that evolution in the past was caused by the NDS *mechanism* of natural selection acting on random variation. Evolution is a "wise-after-the-event" description of the natural world. The diversity of living things may have been the result of mutations or it may not. The mutations themselves may have been the result of chance or they may not. Evidence from present-day studies is inconclusive and evidence from the past (not the changes themselves, but the evidence of what *caused* the changes) is mostly missing. You can *argue* that natural selection caused evolution in the past but you cannot *demonstrate* it. For example, the fossil record provides strong evidence that birds evolved from small dinosaurs, but the various proposed ways in which this might have happened by natural selection are simply Just So stories. The countless small daily discoveries in laboratories and elsewhere that confirm the underlying relationships between all living organisms are evidence for evolution, but not necessarily for Darwinian evolution. Unless the popular accounts of evo devo are misleading, the development of new structures like wings or flowers during evolution was not the result of new mutations, but of old genes being used in a different way.

Creationists have always resisted evolution because it is cruel, wasteful and chancy. Those things may be true of the NDS, but evolution as it actually happened may not have been like that at all. Non-Darwinian evolution may be no more chancy, wasteful or cruel than growth, or the development of the embryo.

Non-YEC creation

If it is important in evolutionary biology to hold loosely to interpretations but tightly to facts, it is important when studying the Scriptures to stick to the actual words of the Bible. There is nothing written in the actual text of Genesis about the age of the earth, nothing in the Curse (of Genesis 3) about the death of animals and nothing anywhere about dinosaurs or flood geology. YEC publications are prone to mingle events and ideas which are not in the Bible with others which are. Just as a modern TV programme on dinosaurs will depict them as variously coloured, and making squeaking or booming noises (totally imaginary, because all we have got is their bones) a YE creationist publication will explain how dinosaurs were housed on the ark, or describe massive volcanic activity during or just after Noah's Flood. These extra bits and pieces are *interpretations*: interesting and ingenious, but not part of science (in the case of the TV programmes) and not part of Scripture (in the case of flood geology). Many of the disagreements and difficulties are not to do with the actual facts of nature, or the actual words of Scripture, but with Just So stories based on them. These have changed a lot in the last half-century: how do we know they will not soon be quite different again?

Where the stories from the Two Books overlap, there is general agreement between them. The creation of the universe, and of the world and the living things found in it, is described as a series of episodes, roughly in the same order, and culminating in the appearance of mankind. It is a mistake to try to force a close correlation between the Scriptural and scientific stories; it goes against the principle of complementarity and there are insufficient data. (And look at what happened to poor Philip Henry Gosse, when he tried!)

Two big difficulties, even in this centre ground, are to do with time.

The age of the earth.

The problem here is the *huge* disparity in possible ages: a few thousand years (YECs) against many billions of years (everyone else); roughly a million-fold difference. My hunch is that the "scientific" age is probably too long and the YEC age is far too short - but I would say that, wouldn't I? After all, both views are counter-intuitive. Surely, if the world really was billions of years old, it would have worn completely flat hundreds of times over, and all would now be gravel, dust and mud. And if the world really was a few thousand years old there's simply no way that all those tens of thousands of feet of chalk and limestone could have formed, let alone all the rest of the sedimentary rocks (and that's just to start with). Estimates of past time are indirect: you cannot measure past time, you can only measure *present amounts* (of metres of sediments or mass of the products of radioactive decay), and the interpretation of these depends on the assumptions that you bring to the measurements. However the scientific datings have the enormous advantage of consilience: measurements from all the different sciences broadly agree. Young earth creationists claim that the "millions of years" measurements are wrong because any datings that do not fit in with current ideas about the age of the earth are simply ignored or discarded as "discrepant" and that only ages that support evolutionist theory are allowed to stand. Evolutionists will reply that the creationists cannot talk because they do not base their dates on scientific findings at all, but on a commitment to their particular interpretation of the early chapters of Genesis.

There is no way of reconciling these ages as they stand, but here are two observations. Firstly, although the ages are absurdly different, this is not something that troubles the *imagination*. We cannot imagine a billion years, but neither can we imagine ten thousand years. It is like the numbers in astronomy: if someone added or removed a few noughts most of us would hardly notice. Few people can give the distance in light years of the Crab Nebula or the dates of the Ordovician period, and it is difficult to remain indignant about something which is both hard to remember and opaque to the imagination. Secondly, think back to the old Galileo problem. Imagine two of his contemporaries, A and B, having a discussion. Let us suppose that they both know that the sun is millions of miles away. A claims that the earth orbits the sun, and B says, no, the sun goes round the earth. After a burst of mental arithmetic, B announces that the sun, going from East to West in a day, must be travelling at several million miles per hour, whereas A reckons its speed as 0 mph. Here is another million-fold difference. And yet a simple change of theory removes the numerical problem immediately. Perhaps there is a discovery awaiting us that will do the same for the huge discrepancy in estimates of the earth's age.

Evolutionists are quite sure that the world is many billions of years old and YECs are equally sure that the world is only a few thousands of years old. Most scientists, but not all, are in the first group. Whatever the truth of the matter, there is one thing we must insist on. Mere time cannot allow something to happen that is intrinsically impossible. However long the passage of time, something cannot drift from one category to another. So, over time, inanimate matter does not just *become* alive, nor matter become mind; that which is irrational does not just become rational, and random noise does not become information.

The six days of Creation.

The Scriptures refer to Creation taking place over the course of six days and Darwinism flatly disagrees. This is a "Newton's dog" problem, with the added difficulty (common in religious disagreements) that it is of great significance to some people and none whatsoever to others. Further complications are that in the Genesis account the created world was there *before* the first day and that the seventh day had no recorded ending. [3] Also, although the creation events in the first chapter of Genesis are global and universal, the standpoint is local when it comes to defining the days of creation. "Evening" and "morning" depend on where you are. When it is daytime in the Near East it is night on the other side of the world. [4] This is further confirmation that the Genesis account is a story being told. I also favour the idea, which I found somewhere in David Pawson's extensive writing about Creation, that the six days are of *theological* importance, rather than historical or scientific. They are partly to show that Creation is "all in a week's work" for God and partly to set the scene for the human / divine drama that is to come.

Incidentally, the time divisions of our lives mostly depend on astronomical events. A day is the period of rotation of the earth, a month is related to the period of revolution of the moon round the earth, and a year is the period of revolution of the earth round the sun. But there is nothing in astronomy that corresponds to a week; especially with the pattern of six days of work followed by one day of rest. Here there is food for thought: what is the exact nature of the week and the relationship between the seven-day week, the Genesis account of Creation, and the fourth commandment (to keep the Sabbath Day special) ?

Trees and lawns

There is one area of discussion where the two sides are drawing slightly closer together. Evolutionists, from Darwin onwards, have pictured living things as branches and twigs on a *tree* of life. Creationists, disbelieving in common descent, pictured instead a *lawn* of separately-created species. Both diagrams have now been modified. The evolutionary trees have been redrawn to look more like bushes. Creationists meanwhile have accepted that microevolution has caused much branching and that many lines have become extinct; so their lawns have become 'orchards.'

At the base of the diagrams there is disagreement (It happens to be the region of the diagram about which nothing is known. Evolutionists do not know how the phyla are related to each other, and creationists do not know how God created.) But the readiness of each group to modify their original concepts is encouraging.

Notes on Chapter 17

1. From *The New Biology* (1987) by Robert Augros and George Stanciu

2. From the Introduction to *The Origin of Species* (1956) by W R Thompson

3. Genesis 1: 1 to 5 refers to the creation of the world; Genesis 2: 2 and 3 refer to the seventh day. Also see *Clues to Creation in Genesis* by P J Wiseman (but his conclusions will not be welcomed by creationists)

4. See *Seven Days that Divide the World* (2011), Appendix E, by John Lennox

Chapter 18

Advice to Students

This is what I would say to students if I was still teaching:

Check statements against the evidence

When you hear or read controversial statements, apply the Nuffield filter: ask yourself "Is this certainly true . . . ?" and so on. This particularly applies when the author or speaker has an axe to grind, or is described as a "something-ist" (feminist, Marxist, and so on). "Isms" are not always wrong, but their followers, by definition, will be looking at the world in a particular way. Their statements may be selective or biased, and they are trying to win you to their position.

This is particularly important with statements about the remote past, many of which are sheer guesswork. Take them lightly.

Beware of the word "scientific" when used as a point-winner in argument. I have just read a book which attacks all Christian views on evolution, and the word "scientific" is used throughout to support the views the authors approve of. In every case it is *historical* science in question. Remember that the prestige of science is attached to hard science (testable, present-day science), not to historical science which may be unreliable.

Look beyond the packaging

Be wary of modern books and television programmes which rely on brilliant photography, clever graphics and computer simulations. What you read or watch will have been carefully edited (you don't see the thousands of feet of discarded film). Nature programmes are factual - but the facts themselves will have been selected to tell a particular story. Once, most Nature films and programmes encouraged you to marvel at the beauty and extravagance of life in the wild. Early Disney wildlife films emphasised the cuteness or 'humanness' of animals. But many modern television films major on the nastiness of nature, with shots of animals killing and eating each other. C S Lewis has pointed out that you can illustrate almost anything from nature, which "has many different phenomena in stock."

Gaps in the story being presented, particularly in astronomy and evolution, are filled with computer-generated images. Remember that something does not become true because it is cleverly portrayed with state-of-the-art visuals. A computer simulation is just that, a *simulation*, a product of the imagination.

Be suspicious of *reconstructions*, for example those family groups of "cavemen" found in museum exhibits and book-illustrations. They are based on flimsy evidence, perhaps just a single tooth or a few fragments of skeleton.

This sounds very negative and suspicious! But scientists are *meant* to question everything.

Use your imagination as well as logic.

The theory of evolution and the Genesis account are both Creation Stories, and *stories* are apprehended by the imagination as well as the intellect. Try some lateral thinking. Ask of everything, "Is there another way of looking at this?" A local art gallery here in Somerset recently held an exhibition of abstract art and the curators hung one important painting upside down. No one realised until the artist himself came to see the show and pointed out the mistake. He wasn't pleased; but there, that's abstract art for you. Try holding *ideas* upside down, and see what they look like then. Did species A turn into species B by gaining some segments, or did B turn into A by losing some? If all you had were the bones, how might you demonstrate to someone that a penguin is not transitional between a fish and a seagull? We have all heard about how the giraffe got its neck: why don't we hear about how the mouse got its neck? Did your dog *evolve* its tail?

Don't be overawed by the numbers

All those billions of galaxies and millions of years of evolution can overwhelm the imagination. But they are just numbers. I was in a "rock shop" the other day, looking at the minerals and fossils which were for sale. The card on an ammonite stated "*Androgynoceras* - Lower Lias - 180 million years old" It was a nice fossil - but it wasn't just the ammonite that was 180 million years old. So were the stones that the shop was built with and the pebbles in the garden behind, and even much of the dust on the floor! These huge figures are partly a matter of scale and unit. 180 million is a huge figure: but I have just worked out that my friend's house in Birmingham is approximately 180 million millimetres away from our house in Somerset. Note, too, that we humans are nicely in the middle of these sizes: our bodies are billions of times smaller than galaxies and billions of times larger than sub-atomic particles.

"Scientific" attacks on faith

This is for Christian (or "religious") young readers who might be troubled that the evolution debate is being used in an attempt to undermine their beliefs or relationship with God. Do not be dismayed when atheists call your faith deluded. They are bound to say that, aren't they? What does a man who lives in darkness know about light? You would not go to a Swiss farmer to find out about the Pacific Ocean, and there's no point asking an atheist if you want to know about God. There has been a spate of atheist books recently and unfortunately some of them have been associated with science. Christian readers should not be discouraged because firstly, such attacks are to be expected, and secondly, they have happened for centuries. They are nothing to worry about.

Every few years - roughly once a decade - there is a notable attack on Christianity, usually associated with a particular individual or group. A few years later on that individual is forgotten, the ripples have moved away, and Christendom is pretty much as it was before. Yes, church numbers go up and down, but they always have done; and anyway the spiritual strength of something is seldom a matter of numbers. Each attacker thinks to himself that "That's it! Christianity will never survive my onslaught!" One of the best-known examples is that of Voltaire, who wrote "Within 100 years of my death Christianity will be swept from existence and pass into history." He died in 1778 and his house was later bought by the Geneva Bible Society.

Attacks and defences are often represented by books coming out in *pairs*; an atheist book followed by its corresponding Christian response. Recently we have had *The God Delusion* with its counterpart *The Dawkins Delusion*, and , just as I was writing this section, another pair has been published, *The Greatest Show on Earth* by Richard Dawkins and *The Greatest Hoax on Earth?* by Jonathan Sarfati (see Bibliography). I'm not sure that these Christian 'response-books' are the best way of dealing with the problem. Most of the irritants just pass away with time. Who bothers nowadays with those apostate bishops of the '60s and '70s? Does anyone still read Haldane or Hogben? The old arguments are recycled with slightly different wording, but it is hard to measure what effect they have.

In the last chapter of his recent book, *Why Us?*, James Le Fanu argues that the current attacks on faith by the ultra-Darwinists is to cover up weaknesses in their own position:

" . . their remorseless hostility to religion ('a dangerous collective delusion') is best interpreted as a rhetorical device, a sleight of hand to distract attention from the intellectual weaknesses of scientific materialism."

As we should all know by now, Christianity thrives under persecution. One of the best ways to make sure that it survives is to try to stifle it, as the recent history of both China and the Soviet Union has shown. This is even true of smaller movements within Christianity. I am surprised that no one has told those who continually attack the creationists with aggressive atheism, that they are actually acting as recruiting-sergeants for their opponents. When I was teaching biology there was nothing at all in the textbooks about any form of creationism. Then the atheists became more vocal and aggressive and the syllabus now includes Intelligent Design.

In one of his books C S Lewis points out that you cannot insult God. He describes the attempt as being like that of a lunatic trying to put out the sun by scribbling the word "darkness" on the wall of his cell. This is worth remembering next time an atheist tries to use science to attack God.

Chapter 19

The Genesis of Nature and the Nature of *Genesis*

I started to write this book in response to Dave's challenge, made over a decade ago. Some arguments from those days have been overtaken by new factual knowledge, especially in evo devo and gene-sequencing, and by new ways of thinking. I have taken no party line: it would be of little help to those like Dave who wanted to know what the different views were. In a sense the views are not very important. If someone came up with proof that the neo-Darwinian synthesis was right it would not affect my Christian beliefs. If there was proof that the earth was much younger and that the history in Genesis is similar to other human history, I would be surprised but not put out. But the whole point is that there *is* no proof of what happened in the past, and never can be. The ultra-Darwinists are mistaken in supposing that evolution indicates atheism, but that has nothing to do with science, and there are better arguments against them than those put forward by the creationists.

You cannot write a final "Conclusion" to a book about an ongoing debate: it is like writing the last chapter of the biography of a living person. Tomorrow's discovery may put everything back in the melting-pot. Science never stands still. Nor does Biblical interpretation: YEC creationism, as a movement, is only a few decades old. But it is unhelpful to raise a topic and then leave it hanging in mid-air. It is good for schoolmasters to ask lots of questions, but we must also point towards answers. If you challenge neo-Darwinism you will be asked "What is your alternative?" When I talked to a lively Christian Youth Group recently and told them about the different positions in the Great Debate, they wanted to know: "What do *you* think?"

My provisional reply is to say "Yes" to evolution and "No" to natural selection as its mechanism. For most scientists and other thoughtful people Young-earth creationism won't do; they find it simplistic and more like preaching than explanation. Problems concerning the age of the earth, the "cousinship" of animals and of plants, the patterns in embryology, the details of stratigraphy and the fossil record: these questions are not adequately addressed. They are not mentioned in the Scriptures, so discussion about them is speculative even from a biblical viewpoint. And neo-Darwinism won't do either. The role of DNA is exaggerated and the role of the living cell downplayed. The origin of beneficial variation and of totally new structures has not been adequately explained. Natural selection is often a superfluous idea, invoked because it is the party line rather than because there is any evidence for it. Some arguments used to support the NDS are special pleading: natural selection has just *got* to be true ("blow the evidence, just agree with me!"). Much of what is confidently claimed about what happened in the past is sheer guesswork; and one scientist's guess is as good as another's.

Human ideas (like the events on an epigenetic landscape) tend to run in channels. Creationists say the same things and Ultra-Darwinists make their assertions, over and over again. Familiar arguments are rehearsed and insults traded, almost in pantomime fashion ("Oh yes it is!" - "Oh no it isn't!"). If you look on the Internet at any interchange of views about a new idea in evolution, the protagonists quickly abandon the fresh details and return to their slanging matches. This is pointless and silly: a fresh approach is needed. ID may be inconclusive, but at least it is different. Evo devo is excitingly new, and has the advantage that it is actually based on research. There have been plenty of fresh ideas about evolution during the last half-century, but they have mostly been squashed by the dead hand of neo-Darwinism, which cannot be empirically tested and is often just asserted dogmatically.

It need not be like that. In my reading I have been struck with how many biologists, and other scientists, are looking beyond the NDS and expecting a paradigm-shift. I put forward the idea of Innate Transformation as a Just So story in Chapter 8; I am not particularly attached to it and will lose no sleep if evidence shows it is wrong. But something like it will turn out to be true: the capacity to evolve in a structured way is built in to living systems; random mutation plays a minor part in evolution; the form of organisms is an interplay between their total chemistry (not just the DNA) and the environment; natural selection is just quality-control. Non-random, innate factors are involved in development; whether embryological or evolutionary. John Waller's book makes it clear that these possibilities are closer to Darwin's own thought than NDS reductionism. This doesn't mean that Darwin was *right*: he somehow enclosed nature in a mechanistic, materialist box which many people find stifling to their imaginations. That is why *The New Biology* by Augros and Stanciu make such stimulating reading and why reviews of *Why Us ?* by James Le Fanu refer to it as "a breath of fresh air."

Darwinists are opposed to such ideas because they smack of "vitalism" (the idea that organisms have some *élan vital* inside them that is beyond scientific investigation) and "teleology" (the idea that living things are there for a purpose). But those criticisms miss the point. Living things *are* alive and we have to face it. The single most important thing about an animal or plant is that it is alive. If you kill it you make an inexorable, irreversible change: you change its state for ever. And organisms actually *do* things. They are active, not passive. They behave in a purposeful way. It is not scientific to deny reality. Why is it so hard for biologists to accept that living organisms are different from things that are not alive? Why is it so hard, for that matter, to accept that evolution has a direction? You can marshal clever arguments to show that a tiger is no more advanced than an amoeba, but does anyone really believe it? Gould wrote a lot about directionality, mostly to the effect that evolution was contingent and that human beings were no more evolved than bacteria; but he also referred to a "vector of progress" in evolution. Darwin, too, seemed to change his mind, finally accepting some form of progress in his scheme of evolution, describing "the production of the higher animals" as "the most exalted object which we are capable of conceiving" in the last paragraph of *The Origin of Species*. Like Gould and Darwin, some TEs seem not quite clear about it either, although most of those I've read believe in randomness in evolution but directionality in God's guidance of it. This question of direction and purpose in evolution has nothing to do with biological science, but is a matter of worldview. Atheists are bound to disbelieve in directionality, whatever the evidence, and Christians and other theists are bound to accept it.

The current situation is in a state of flux. Perhaps it can be summarised in a historical way, like the geological periods: the 19th Century was the age of Darwinism; the 20th Century was the age of neo-Darwinism; the 21st Century will be the age of post-Darwinism. (It would be good if the word "Darwin" - two syllables which send people's thoughts off in wrong directions - was dropped altogether.) Evo devo, and research on the biology of how genes work within the cell, should do for this century what the novel science of genetics did for the twentieth century. Perhaps the latest book by Fodor and Piatelli-Palmarini will shock the scientific Establishment into a reappraisal of natural selection. Perhaps, as Augros and Stanciu have anticipated, a biological revolution equivalent to that which gave relativity and the quantum theory to physics will release the life sciences from the mechanical straitjacket of the NDS. Scientists and philosophers might even catch up with the writer of John's Gospel: "In the beginning was the *Logos* . . ." Most people write and talk as though we are nearing the end of this particular journey of discovery; as though nearly everything is clear and that it's just a matter of taking sides between the two extremes of evolution and creation. We are actually still near the beginning of this road; new discoveries, and wiser thinking, will alter the picture in the years to come.

What did happen "In the Beginning"? What is the genesis of Nature (the origin of species)? No one knows, because it happened in the distant past. Even if there were adequate words to describe what happened, they would mean little to us because it would be outside our experience. Both biology and Scripture have provided *stories* to help our understanding (the former answering some "How?", the latter some "Why?", questions): hence the idea of the Two Books. One Book tells us that God created every thing from stars to plants, fish to mankind, in a series of episodes. The story in Genesis is a précis, told in simple language, which even children can understand. In no way is it a piece of science writing. Like a great work of art it engages the whole personality; the imagination just as much as the logical capacity (a point missed by both literalists and reductionists.) The other Book tells us that the earth is ancient, that living things appeared in succession and that all life is related, using much the same biochemistry. (This "scientific" story is not as complete as people make it sound: nothing is actually known of the origin of life or the early stages of evolution.) These two accounts are compatible but no one knows how they dovetail together. The evidence from the fossil record, the only concrete evidence for what organisms were like in the past, is that animals and plants appeared in sequence. There are few clues as to how they were connected but it looks as if creation was a matter of gradual, stepwise development, like that of a developing embryo. The alternative is for some sort of sudden creation to have occurred, with a corresponding "appearance of age" - what I have called Gossery. No one can prove which is right because there is no way of finding out: the evidence is the same for both views and the results would be identical. Towards the end of both creation stories

Mankind appears. In both Books his physical nature, i.e. his body, is not different in kind from that of the other animals, but in the Scripture account something special happens to him. In Genesis 1 we read that "God created man in his own image." That cannot mean anything physical because the Bible makes it quite clear that God is Spirit, without bodily parts or passions. The commonsense view is that what God has given us here are those things that make us specifically human: our appreciation of goodness (our moral sense), truth (our intellectual ability), beauty (our creativity) and a response to God himself.

Because the two accounts, the Two Books, do not dovetail neatly together, there is room for many different interpretations of the Creation story. That is why it is not helpful to hold a rigid view and to try to put everyone else right. Let us attempt to correct actual mistakes or errors of fact, by all means, but let us sit light to any particular theory.

If all we had was the "Book of Nature" we would be in a sorry case. We would have to settle for a material, pitiless, indifferent universe; the suffocating world of the selfish gene, or the endless dismay of the bleak Wellsian darkness. If you choose to turn your back on the light, yes, you get darkness. If you make out that goodness, truth and beauty are subjective things, mere by-products of evolution, you are almost certain to end up with evil, falsehood and ugliness.

So thank goodness, or rather thank God, that we have that second "Book" as well. Let men of wisdom continue to follow Bacon's advice and "endeavour an endless progress or proficience" in both books.

GLOSSARY

Part 1, contains terms which may be unfamiliar to readers who left Biology behind at school. Here is a short list of such words, with meanings..

Anthropomorphism
>Giving human characteristics to other organisms or objects.

Biochemistry
>The chemistry going on inside living organisms (often involving complex substances such as proteins, hormones and DNA); the chemistry of *organic* molecules.

Chloroplasts
>The tiny green bodies (organelles) inside many plant cells that contain chlorophyll

Convergence
>The 'becoming alike' of two different things. Example: a shark is a fish, a dolphin is a mammal, but they have similar streamlined bodies, well-adapted for living in water.

Cytology
>The study of cells

Cytoplasm
>The "living stuff"; the main component of a cell. (The other part is the nucleus, containing the DNA). Most of the cell's chemistry happens in the cytoplasm.

DDT
>An insecticide, once widely used but now banned. (It harmed other organisms, and some insects became resistant to it.)

DNA
>Deoxyribonucleic acid. Genes (arranged in bodies in the nucleus called chromosomes) are made of DNA. The "genetic material" of all organisms except viruses.

Directional mutation
> Mutations (changes in the DNA) which are not random but which directly cause the cell or organism to become better adapted to its environment.

Ecosystem
> A region of the world defined by its living contents: a self-contained community of plants and animals, plus where they live (such as a coral reef, acid heath, desert or rain-forest).

Empirical evidence
> Evidence based on experiment or observation (not on theory or guess-work).

Epigenetic
> (A tricky word with variable meanings) Usually refers to genetic / inheritable traits or changes that are not caused by the DNA code.

Genome
> The total genetic information of a species.

Genotype
> The particular set of genes of a particular organism (as opposed to its observable features, its Phenotype).

Mitochondria
> Tiny bodies (organelles) in the cell that do respiration, providing the cell with energy (the "batteries" of the cell). Like the chloroplasts, they have small amounts of their own DNA.

Mutation
> 1. A change of any kind. 2. In biology, an inheritable change in the genetic material.

Niche
> (= Ecological niche) Where a species "fits in." The habitat and environment (including other organisms) of a species, plus the particular rôle of that species in that environment.

Nucleic acids
> DNA and RNA. Complex organic molecules (the name refers to the cell nucleus, where DNA is mostly found) DNA is the genetic material of all organisms apart from viruses which have RNA.

Nucleus
> That organelle, centrally placed in a typical cell, which contains the chromosomes, made of DNA.

Organelles
> As organs are to the body, organelles are to the cell: tiny structures in the cytoplasm which perform the various jobs of the cell.

Palaeontology
> The study of fossils

Phenotype
> The observable features of an animal or plant. Things like 'Tall' or 'Albino'; 'female' or 'spotted'. (As opposed to the Genotype, the set of genes.)

Physiology
> The branch of biology which deals with the *working* of the body (things like respiration, digestion, reproduction) as opposed to Anatomy, the *structure* of the body.

Polymorphism
> The condition in which members of a given species have different shapes, colours, &c. resulting from slightly different forms of DNA (for example peas with smooth or wrinkled seeds; coloured or white (albino) peacocks).

Punctuated Equilibrium
> An attempt to account for stasis (organisms staying unchanged for ages) and saltation (sudden jumps) in the fossil record. (Briefly: there were very rapid changes in certain places and at certain times but not generally; these events are only sporadically shown by the fossil record.)

Radiation
> (in evolutionary biology) The divergence of members of a single group into many different forms, adapting to different niches or zones

Replicator
> Something that can make a copy (or copies) of itself

Saprophyte
> A plant (such as a fungus or some orchids) without chlorophyll, that cannot 'feed itself' by photosynthesis, but has to get nutrients from decaying organic matter.

Taxon (Plural: Taxa)
> A general word for a division in the classification (such as genus, order, class); a group of organisms that share the same characteristics.

Tautology
> A way of speaking, or an argument, that just says the same thing twice using different words.

Teleology
> The study of final causes; the belief in direction-giving or purposeful forces. Any ascription of purpose to evolution.

Vascular
> Containing vessels (water-carrying vessels in 'higher' plants; blood vessels in 'higher' animals).

Vitalism
> The ascription of a magical element, or "life force" to living organisms

Bibliography

This bibliography is incomplete. There are few primary sources. I have tried to include all the books referred to in the main text and some others that I recommend. I have included only a few books from the extreme ends of the spectrum (ultra-Darwinists and YE creationists); many cover the same ground over and over again.

ALEXANDER, Denis and WHITE, Robert 2004 *Beyond Belief* Lion
Sub-titled *Science, Faith and Ethical Challenges*. [just a small section in the middle is about evolution]

ALEXANDER, Denis 2008 *Creation or Evolution Do We Have to Choose?* Monarch Books [both *very* scientific and *very* Christian; anti-ID]

ALEXANDER, Denis and SPENCER, Nick 2009 *Rescuing Darwin - God and evolution in Britain today* Theos (a booklet of 65 pages) [up-to-date; full of facts and figures. Over-mention of Darwin]

ASHTON, John F (Editor) 1999 *In Six Days - why 50 Scientists Choose to Believe in Creation* New Holland (Australia) Pty Ltd [all 50 have doctorates; 18 professorships; concise but familiar arguments]

AUGROS, Robert and STANCIU, George. 1987 *The New Biology* New Science
Library [a fresh, outsiders' view]

BEHE, Michael. 1996 *Darwin's Black Box* Free Press (Simon & Schuster)
 2007 *The Edge of Evolution* Free Press (Simon & Schuster)
[a biochemist champions ID]

BERRY, R J. 1996 *God and the Biologist* Apollos [a more general book than most of the ones listed here: about one quarter of it is about evolution]

BLACKMORE, Vernon and PAGE, Andrew. 1989 *Evolution - The Great Debate* Lion [historical approach; good illustrations]

BROCKMAN, John (editor). 2006 *Intelligent Thought* Vintage   [16 essays attacking ID]

BROWN, Andrew. 1999 *The Darwin Wars - How Stupid Genes became Selfish Gods* Simon and Schuster

CARROLL, Sean. 2006 *Endless Forms Most Beautiful* Weiderfield & Nicolson (UK) Subtitled *The New Science of Evo Devo and the Making of the Animal Kingdom.* [excellent: I learned something on almost every page]
   2006 *The Making of the Fittest* Quercus

COLLINS, Francis. 2007 *The Language of God* Pocket Books Subtitled *A Scientist Presents Evidence for Belief.* [the author was head of the Human Genome Project]

COYNE, Jerry. 2009 *Why Evolution is True* Oxford

DARWIN, Charles. 1859 *On the Origin of Species* John Murray, London [what can one say ? It's a remarkably *homely* book, mentioning the London Pigeon Clubs, Mr. Bakewell's flock of sheep and Dr. Cruger's 'bucket-orchid'. There are several hundred pages of minute details and if it wasn't for the summaries at the end of each of the 25 chapters, it would be hard to see the wood for the trees]

DAWKINS, Richard. 1986 *The Blind Watchmaker* Longmans [a modern account of neo-Darwinism. There are some good things in it but it's hard to trust the author, who writes as an atheist who happens to be a zoologist, rather than the other way round. He gives the game away in the introduction, stating that it isn't enough just to give the evidence: you have to be an advocate and "use all the tricks of the advocate's trade." I doubt if readers want to be tricked]

    2005 *The Ancestor's Tale* Phoenix [scholarly, enjoyable, but quite hard going]

    2009 *The Greatest Show on Earth* Bantam press

DE BEER, Sir Gavin 1940 *Embryos and Ancestors* (reprinted 1971 - Oxford)
    1964 *Atlas of Evolution* Nelson
    1971 *Homology, an Unsolved Problem* Oxford Biology Readers, No. 11

DENTON, Michael. 1985 *Evolution- A Theory in Crisis* Burnett Books [I recommend this book highly]

DOVER, Gabriel 2000 '*Anti-Dawkins*' an article in *Alas Poor Darwin*; see under Rose, Hilary and Stephen. [short, but eye-opening]

FODOR, Jerry and PIATELLI-PALMARINI, Massimo 2010
What Darwin Got Wrong Profile Books [important, recent book by two atheists attacking the NDT]

FOSTER, Charles 2009 *The Selfless Gene* Hodder and Stoughton

GOULD, Stephen Jay 1977 *Ontogeny and Phylogeny* Harvard University Press

    1978 *Ever since Darwin*
    1981 *The Panda's Thumb*
    1983 *Hen's Teeth and Horse's Toes*
    1985 *The Flamingo's Smile*

1993 *Eight Little Piggies*
1996 *Dinosaur in a Haystack*
[six collections of entertaining eassays on all aspects of palaeontogy and evolution]
2001 *Rocks of Ages - Science and Religion in the Fullness of Life* Jonathan Cape

HAYWARD, Alan. 1985 *Creation and Evolution* Triangle Books [an OEC author; he argues against both evolution and Flood Geology]

HITCHING, Francis. 1982 *The Neck of the Giraffe* Pan Books
The sub-title of this book is *Where Darwin went wrong*.

HOYLE, Fred. 1983 *The Intelligent Universe* Michael Joseph
Subtitle: *A new view of creation and evolution*.

KIRSCHNER, Marc W. & GERHART, John C. 2005 *The Plausibility of Life* Yale University Press Subtitled *Resolving Darwin's Dilemma*. [Our local library could find no copy of this wordy book for me in Somerset, but sourced one from a *prison* in a distant county. I now have a splendid mental picture of a group of prisoners, like Fletcher and Godber in *Porridge* discussing whether deconstrained processes can alter the phenotype without resulting in lethality]

JOHNSON, Phillip. 1993 *Darwin on Trial* (2nd Edition) InterVarsity Press [the best of Johnson's books, and the best ID book I have read]
1995 *Reason in the Balance* InterVarsity Press
1998 *Objections Sustained* InterVarsity Press

LEAKEY, Richard E 1979 *The Illustrated Origin of Species by Charles Darwin* Faber [this book is *The Origin* abridged and illustrated by Leakey]

LE FANU, James 2009 *Why Us ?* Harper*Press*
There is a subtitle: *How Science Rediscovered the Mystery of Ourselves.*

LENNOX, John C 2007 *God's Undertaker* Lion subtitled *Has Science Buried God?* [excellent: I would have recommended this to Dave if it had been available]
2011 *Seven Days that Divide the World* Zondervan Subtitled *The Beginning According to Genesis and Science*

LEWIS, C S 1940 *The Problem of Pain* Geoffrey Bles [Ch 5 has reference to evolution]

LEWONTIN, R C 1993 *The Doctrine of DNA* Penguin
Subtitled *Biology as Ideology*, - its original title.

LOGAN, Kevin. 2002 *Responding to the Challenge of Evolution* Kingsway

MAYR, Ernst. 2002 *What Evolution Is* Weidenfield & Nicolson [solid, worthy book by one of the founders of the NDS]

MEDAWAR, P B 1967 *The Art of the Soluble* Methuen (Pelican 1969)
[a book of essays on science and biology, several on 'evolutionary' topics]

MEDAWAR, Peter and Jean. 1977 *The Life Science* Wildwood House

MEYER, Stephen C., with four co-authors 2007 (2009, U.K.) *Explore Evolution* Hill House Publishers; subtitled

The Arguments for and Against Neo-Darwinism [this book is arranged, and intended, as a school textbook]

MILLER, Kenneth R 1999 *Finding Darwin's God* HarperCollins, New York subtitled *A Scientist's search for Common Ground between God and Evolution*

MILTON, Richard 1993 *The Facts of Life: Shattering the Myths of Darwinism* Corgi

MORELAND, J P and REYNOLDS, John Mark 1999 *Three Views on Creation and Evolution* Zondervan [a multi-authored book in the form of discussions]

MORRIS, Henry and WHITCOMB, John 1969 *The Genesis Flood* The Evangelical Press

NUMBERS, Ronald L 1992 *The Creationists - The Evolution of Scientific Creationism* University of California Press

PATTERSON, Colin 1978 *Evolution* British Museum (Natural History)

PENNOCK, Robert T 1999 *Tower of Babel - The Evidence Against the New Creationism* Bradford Book MIT Press (Cambridge, Mass)

POLKINGHORNE, John. 1986 *One World* SPCK subtitled *the Interaction of Science and Theology*

RIDLEY, Mark. 1985 *The Problems of Evolution* OUP

ROSE, Hilary and Stephen (editors). 2001 *Alas Poor Darwin - Arguments Against Evolutionary Psychology* Vintage [A collection of articles by distinguished authors including the two Roses themselves, Gabriel Dover, Mary Midgley and Stephen Jay Gould: essential reading for any student of evolution]

SARFATI, Jonathan. 1999 *Refuting Evolution* Master Books
 [a response to *Teaching about Evolution and the Nature of Science*, published in 1998 and distributed to teachers throughout America; for me, the best of the YEC books]
 2010 *The Greatest Hoax on Earth?* Creation Book Publishers subtitled *Refuting Dawkins on Evolution*

SPETNER, Lee. 1998 *Not by Chance!* Judaica Press [a scientific and mathematical attack on neo-Darwinism]

STROBEL, Lee. 2004 *The Case for a Creator* Zondervan

SWITEK, Brian. 2011 *Written in Stone* Icon Books subtitled *and the Story of Life on Earth*

THOMPSON, D'Arcy. 1961 *On Growth and Form* Cambridge [the abridged edition; the original book was written in 1917]

THOMPSON, W R, Professor, FRS 1956 *Introduction to "The Origin of Species"* Everyman
Library No. 811

WALLER, John 2002 *Fabulous Science* Oxford subtitled *Fact and fiction in the History of Scientific Discovery.*

WILSON, Edward O. 2000 *The Diversity of Life* Penguin

WISEMAN, P J. 1976 *Clues to Creation in Genesis* Marshall, Morgan & Scott [this book consists of two earlier books (from 1936 and 1946) put together and edited by Donald Wiseman, the author's son]

WOODWARD, Thomas. 2003 *Doubts about Darwin* Baker Books [a support of ID from the standpoint of the *rhetoric* of the debate]

INDEX

Acworth, Bernard, 17, 130*n*
Adaptive Radiation, 32, 99
Ambulocetus, 156
Archaeopteryx, 32, 95, 109, 150, 157
Augros, Robert (and George Stanciu), 53, 127, 158, 212, 227*f*
Australopithecus, 156

Bacon, Francis, 120, 128
Behe, Michael, 85, 100, 117, 141*f*, 144
Bellamy, David, 108
Berry, R.J., 143*f*
Blackmore, Vernon (and Andrew Page), 13
Blair, Tony, 128
Blyth, Edward, 186
Brockman, John, 129*n*, 143
Bronowski, J., 97
Bryan, Jennings, 200

Cairns-Smith, A.G., 100
Calvin, John, 179
Cambrian Explosion, 66, 83, 93, 151, 153
Carroll, Sean, 81, 83, 111
Charnia, 74
Chesterton, G.K., 23, 126, 201
Collins, Francis, 129*n*, 198*f*
Complementarity, 144*f*, 212
Consilience, 183*f*
Convergent evolution, 94

Crick, Francis (and James Watson), 58, 108

Darwin, Charles, 1, 5*f*, 9, 11, 17, 25, 30, 32, 39, 40*f*, 45, 50*f*, 62*f*, 67, 76, 85, 97*f*, 108, 112, 120, 139, 150, 162, 185*f*, 191, 193, 204, 227*f*
Darwin, Francis, 187
Davidman, Joy, 126,
Dawkins, Richard, 65, 67, 146, 190, 223
Days of Revelation, 133
Denton, Michael, 49, 52, 140
Desmond, Adrian (and James Moore), 186
Diarthrognathus, 154
Dodd, Ken, 73
Dover, Gabriel, 56, 79, 105*n*

Einstein, Albert, 185
Eldredge, Niles, 51
Eohippus, 155
Epigenetic landscape, 96
Equus, 155
Evolution Protest Movement, 15, 17, 64, 132
Eusthenopteron, 154
Exploratory processes, 98

Flood Geology, 65, 136*f*, 143, 153, 158, 213, 215
Fodor, Jerry (and Massimo Piatelli-Palmarini), 51, 193, 228
Foster, Charles, 146

Framework Theory, the, 133
Freud, Sigmund, 59, 188

Galileo, 127, 199, 217
Genesis Flood, The, 64, 65, 69*n*, 132*f*
Gerhart, John (and Marc Kirschner), 82, 84*f*, 111
God-of-the-gaps, the, 142, 145, 206
Gosse, Edmund, 162
Gosse, Philip, Ch. 13, 160*f*, 215
Gould, Stephen Jay, 51, 57, 65, 68, 93, 120, 131, 141, 160*f*, 199, 207*f*, 227
Grassé, Pierre, 9, 43, 150

Haeckel, Ernst, 30, 37*n*, 88, 107
Haldane, J.B.S., 9, 15, 100, 152*f*, 189, 202*f*, 209
Hayward, Alan, 164
Hebron School, India, 22
HeLa cells, 79
Hitching, Francis, 48, 52
Hoatzin, 157
Homo (Genus), 156
Horse series (of fossils), 31, 155
Hox genes, 81, 83, 92
Hoyle, Sir Fred, 49, 157, 186
Huxley, Julian, 15, 47*n*, 62*f*, 187
Huxley, T.H., 41, 50, 85, 119, 137, 176, 199

Innate Transformation, Ch. 8, 88, 111
Innes, Hammond, 77

Intelligent Design, 68*f*, 122, 124, 139*f*, 198, 224, 226
Intermediate fossils, 154
Irreducible Complexity, 141

Java Man, 156
Johnson, Phillip, 54, 131, 140, 149, 191*f*, 200*f*, 205

Kipling, Rudyard, 107*f*
Kirschner, Marc (and John Gerhart), 82, 84*f*, 111

Lamarck, Jean-Baptiste de, 38*f*, 61, 111, 186
Last-Thursdayism, 165
Le Fanu, James, 6, 77, 127, 167*n*, 188, 198, 223, 227
Lego models, 82, 89, 91
Lennox, John, 129*n*, 145, 204
Lewis, C.S., 76, 130, 139, 170, 182*n*, 184
Lewontin, R.C., 66, 198
Lyell, Charles, 186

Malthus, Rev Thomas, 53, 186
Marx, Karl, 11, 59, 188
Matthew, Patrick, 186
Maynard Smith, John, 51
Mayr, Ernst, 6, 47*n*, 71, 100
Medawar, Sir Peter, 75, 103, 107
Mendel, Gregor, 39, 45, 108, 119, 187
Miller, Kenneth, 100,
Miller, Stanley, 77

Millfield School, 10, 18, 22, *25f*, 31, 58, *71f*, 109, 155, *199f*
Monkton Combe School, *14f*
Moore, James (and Adrian Desmond), 186
Morris, Desmond, 51
Mosaic evolution, 95, 157

Natural Selection, 4, 24, *40f*, Ch. 4, *53f*, 84, 86, 110, 186, *193f*, *226f*
Nebraska Man, 156, 200
Neo-Darwinian Synthesis, The (NDS), *45f*, Ch. 4, *82f*, 85, 90, *94f*, 101, *150f*, 194, *213f*, *225f*
New Scientist magazine, 65, 154, 187, 193
Newton, Sir Isaac, 145, 155
"Newton's Dog" problems, 11, 179, 208, 217
Noah, 65, 123, 134, 136, 152, 175
NOMA, 131
"Nuffield Filter", the, 10, 118, 220

Old Earth Creationism (OEC), 122, *132f*
"Omphalos", *160f*
"*Origin of Species, The*", 17, 21, 32, *39f*, 45, 50, *60n*, 62, 97, 108, 110, 120, 129, 150, 156, 161, 186, 187, 191, 207, 213
Orohippus, 155
Orthogenesis, 92, 101

Page, Andrew (and Vernon Blackmore), 13
Pakicetus, 156
Pasteur, Louis, 78, 205
Patterson, Colin, 155
Pawson, David, 218
Pax 6 Gene, 81, 83, 92
Pekin Man, 156
Pentadactyl limb, 27
Peppered moth, 18, 24, 35, 42, 56
Piatelli-Palmarini, Massimo (and Jerry Fodor), 51, 193, 228
Piltdown Man, 156, 200
Pius XII, Pope, 183
Pliohippus, 155
Polkinghorne, John, *144f*, 204
Pre-Cambrian fossils, 74
Provine, William, 66

Ransome, Arthur, 59
Recapitulation, 29, *37n*, 89, 107
Redi, Francesco, 205
Reductionism, *125f*, *129n*, 189, 204
Ridley, Mark, *37n*, 150
Russell, Bertrand, 167

Saltation, 50, 85, 93
Sarfati, Jonathan, 223
Sayers, Dorothy, *161f*
Scientific American magazine, 131, 141
Scopes Trial, The, 13, 200
Seymouria, 154
Sitwell, Edith, 205
Shakespeare, 125, 144
Spectrum of views, The, *122f*

Spencer, Herbert, 62, 103
Spetner, Lee, 53, 105*n*
Stanciu, George (and Robert Augros), 53, 127, 157, 212, 227*f*
'Survival of the Fittest', 41, 62, 138
Switek, Brian, 101

Taylor, Ken, 64
Theistic Evolution, 121*f*, 143*f*, 212
Thompson, W.R., 50, 60, 110, 207, 213
Thrinaxodon, 154
Transitional fossils, 32, 57, 154
Toynbee, Philip, 203

Ussher, Archbishop, 134

Venter, Craig, 78
'Vertical History', 173*f*

Voltaire, 223

Waddington, Conrad, 75, 96
Wallace, Alfred Russell, 40, 186, 204
Waller, John, 210*n*, 227
Watson, James (and Francis Crick), 58, 108,
Wegener, Alfred, 109
Wells, H.G., 165
White, A.J.Monty, 164
Wilberforce, Samuel, 199*f*
Wilson, Edward, 111
Wiseman, P.J., 219*n*
Woodward, Thomas, 140

Yancey, Philip, 181*n*
Young Earth Creationism (YEC), 64*f*, 68, 122, 124, 127, 132, 133*f*, 152*f*, `83, 190, 225

www.ingramcontent.com/pod-product-compliance
Lightning Source LLC
Chambersburg PA
CBHW071633220526
45469CB00002B/601